# Alfred Russel Wallace
# Letters and Reminiscences
# (Vol

James Marchant

Alfred Russel Wallace

Alpha Editions

This edition published in 2024

ISBN : 9789366385945

Design and Setting By
**Alpha Editions**
www.alphaedis.com
Email - info@alphaedis.com

As per information held with us this book is in Public Domain.
This book is a reproduction of an important historical work. Alpha Editions uses the best technology to reproduce historical work in the same manner it was first published to preserve its original nature. Any marks or number seen are left intentionally to preserve its true form.

(VOLUME 2)

# PART III

## I.—Wallace's Works on Biology and Geographical Distribution

> "I have long recognised how much clearer and deeper your insight into matters is than mine."
>
> "I sometimes marvel how truth progresses, so difficult is it for one man to convince another, unless his mind is vacant."
>
> "I grieve to differ from you, and it actually terrifies me, and makes me constantly distrust myself. I fear we shall never quite understand each other."
>
> —DARWIN TO WALLACE.

During the period covered by the reception, exposition, and gradual acceptance of the theory of Natural Selection, both Wallace and Darwin were much occupied with closely allied scientific work.

The publication in 1859 of the "Origin of Species"[1] marked a distinct period in the course of Darwin's scientific labours; his previous publications had, in a measure, prepared the way for this, and those which immediately followed were branches growing out from the main line of thought and argument contained in the "Origin," an overflow of the "mass of facts" patiently gathered during the preceding years. With Wallace, the end of the first period of his literary work was completed by the publication of his two large volumes on "The Geographical Distribution of Animals," towards which all his previous thought and writings had tended, and from which, again, came other valuable works leading up to the publication of "Darwinism" (1889).

It will be remembered that Darwin and Wallace, on their respective returns to England, after many years spent in journeyings by land and sea and in laborious research, found the first few months fully occupied in going over their large and varied collections, sorting and arranging with scrupulous care the rare specimens they had taken, and in discovering the right men to name and classify them into correct groups.

At this point it will be useful to arrange Darwin's writings under three heads, namely: (1) His zoological and geological books, including "The Voyage of the *Beagle*" (published in 1839), "Coral Reefs" (1842), and "Geological Observations on South America" (1846). In this year he also began his work on Barnacles, which was published in 1854; and in addition to the steady work on the "Origin of Species" from 1837 onwards, his observations on "Earthworms," not published until 1881, formed a distinct phase of his study during the whole of these years (1839-59). (2) As a natural sequence we have "Variations of Animals and Plants under Domestication" (1868), "The Descent of Man" (1871), and "The Expression of the Emotions" (1872). (3) What may be termed his botanical works, largely influenced by his evolutionary ideas, which include "The Fertilisation of Orchids" (1862), "Movements and Habits of Climbing Plants" (1875), "Insectivorous Plants" (1876), "The Different Forms of Flowers and Plants of the same Species" (1877), and "The Power of Movement in Plants" (1880).

A different order, equally characteristic, is discovered in Wallace's writings, and it is to be noted that while Darwin devoted himself entirely to scientific subjects, Wallace diverged at intervals from natural science to what may be termed the scientific consideration of social conditions, in addition to his researches into spiritualistic phenomena.

The many enticing interests arising out of the classifying of his birds and insects led Wallace to the conclusion that it would be best to postpone the writing of his book on the Malay Archipelago until he could embody in it the more generally important results derived from the detailed study of certain portions of his collections. Thus it was not until seven years later (1869) that this complete sketch of his travels "from the point of view of the philosophic naturalist" appeared.

Between 1862 and 1867 he wrote a number of articles which were published in various journals and magazines, and he read some important papers before the Linnean, Entomological, and other learned Societies. These included several on physical and zoological geography; six on questions of anthropology; and five or six dealing with special applications of Natural Selection. As these papers "discussed matters of considerable interest and novelty," such a summary of them may be given as will serve to indicate their value to natural science.

The first of them, read before the Zoological Society in January, 1863, gave some detailed information about his collection of birds brought from Buru. In this he showed that the island was originally one of the Moluccan group, as every bird found there which was not widely distributed was either identical with or closely allied to Moluccan species, while none had special affinities with Celebes. It was clear, then, that this island formed the most westerly outlier of the Moluccan group.

The next paper of importance, read before the same Society in November (1863), was on the birds of the chain of islands extending from Lombok to the great island of Timor. This included a list of 186 species of birds, of which twenty-nine were altogether new. A special feature of the paper was that it enabled him to mark out precisely the boundary line between the Indian and Australian zoological regions, and to trace the derivation of the rather peculiar fauna of these islands, partly from Australia and partly from the Moluccas, but with a strong recent migration of Javanese species due to the very narrow straits separating most of the islands from each other. In "My Life" some interesting tables are given to illustrate how the two streams of immigration entered these islands, and further that "as its geological structure shows ... Timor is the older island and received immigrants from Australia at a period when, probably, Lombok and Flores had not come into existence or were unhabitable.... We can," he says, "feel confident that Timor has not been connected with Australia, because it has none of the peculiar Australian mammalia, and also because many of the commonest and most widespread groups of Australian birds are entirely wanting."[2]

Two other papers, dealing with parrots and pigeons respectively (1864-5), were thought by Wallace himself to be among the most important of his studies of geographical distribution. Writing of them he says: "These peculiarities of distribution and coloration in two such very diverse groups of birds interested me greatly, and I endeavoured to explain them in accordance with the laws of Natural Selection."

In March, 1864, having begun to make a special study of his collection of butterflies, he prepared a paper for the Linnean Society on "The Malayan Papilionidæ, as illustrating the Theory of Natural Selection." The introductory portion of this paper appeared in the first edition of his volume entitled "Contributions to the Theory of Natural Selection" (1870), but it was omitted in later editions as being too technical for the general reader. From certain remarks found here and there, both in "My Life" and other works, butterflies would appear to have had a special charm and attraction for Wallace. Their varied and gorgeous colourings were a ceaseless delight to his eye, and when describing them one feels the sense of pleasure which this gave him, together with the recollection of the far-off haunts in which he had first discovered them.

This series of papers on birds and insects, with others on the physical geography of the Archipelago and its various races of man, furnished all the necessary materials for the general sketch of the natural history of these islands, and the many problems arising therefrom, which made the "Malay Archipelago" the most popular of his books. In addition to his own personal knowledge, however, some interesting comparisons are drawn between the accounts given by early explorers and the impressions left on his own mind by the same places and people. On the publication of this work, in 1869, extensive and highly appreciative reviews appeared in all the leading papers and journals, and to-day it is still looked upon as one of the most trustworthy and informative books of travel.

When the "Malay Archipelago" was in progress, a lengthy article on "Geological Climates and the Origin of Species" (which formed the foundation for "Island Life" twelve years later) appeared in the *Quarterly Review* (April, 1869). Several references in this to the "Principles of Geology"—Sir Charles Lyell's great work—gave much satisfaction both to Lyell and to Darwin. The underlying argument was a combination of the views held by Sir Charles Lyell and Mr. Croll respectively in relation to the glacial epoch, and the great effect of changed distribution of sea and land, or of differences of altitude, and how by combining the two a better explanation could be arrived at than by accepting each theory on its own basis.

His next publication of importance was the volume entitled "Contributions to the Theory of Natural Selection," consisting of ten essays (all of which had previously appeared in various periodicals) arranged in the following order:

1. On the Law which has regulated the Introduction of New Species.

2. On the Tendency of Varieties to depart indefinitely from the Original Type.

3. Mimicry, and other Protective Resemblances among Animals.

4. The Malayan Papilionidæ.

5. Instinct in Man and Animals.

6. The Philosophy of Birds' Nests.

7. A Theory of Birds' Nests.

8. Creation by Law.

9. The Development of Human Races under the Law of Natural Selection.

10. The Limits of Natural Selection as applied to Man.

His reasons for publishing this work were, first, that the first two papers of the series had gained him the reputation of being an originator of the theory of Natural Selection, and, secondly, that there were a few important points relating to the origin of life and consciousness and the mental and moral qualities of man and other views on which he entirely differed from Darwin.

Though in later years Wallace's convictions developed considerably with regard to the spiritual aspect of man's nature, he never deviated from the ideas laid down in these essays. Only a very brief outline must suffice to convey some of the most important points.

In the childhood of the human race, he believed, Natural Selection would operate mainly on man's body, but in later periods upon the mind. Hence it would happen that the physical forms of the different races were early fixed in a permanent manner. Sharper claws, stronger muscles, swifter feet and tougher hides determine the survival value of lower animals. With man, however, the finer intellect, the readier adaptability to environment, the greater susceptibility to improvement, and the elastic capacity for co-ordination, were the qualities which determined his career. Tribes which are weak in these qualities give way and perish before tribes which are strong in them, whatever advantages the former may possess in physical structure. The finest savage has always succumbed before the advance of civilisation. "The Red Indian goes down before the white man, and the New Zealander vanishes in presence of the English settler." Nature, careless in this stage of evolution about the body, selects for survival those varieties of mankind which excel in mental qualities. Hence it has happened that the physical characteristics of the different races, once fixed in very early prehistoric times, have never greatly varied. They have passed out of the range of Natural Selection because they have become comparatively unimportant in the struggle for existence.

After going into considerable detail of organic and physical development, he says: "The inference I would draw from this class of phenomena is, that a superior intelligence has guided the development of man in a definite direction, and for a special purpose, just as man guides the development of many animal and vegetable forms." Thus he foreshadows the conclusion, to be more fully developed in "The World of Life" (1910), of an over-ruling God, of the spiritual nature of man, and of the other world of spiritual beings.

An essay that excited special attention was that on Mimicry. The two on Birds' Nests brought forth some rather heated correspondence from amateur naturalists, to which Wallace replied either by adducing confirmation of the facts stated, or by thanking them for the information they had given him.

With reference to the paper on Mimicry, it is interesting to note that the hypothesis therein adopted was first suggested by H.W. Bates, Wallace's friend and fellow-traveller in South America. The essay under this title dealt with the subject in a most fascinating manner, and was probably the first to arouse widespread interest in this aspect of natural science.

The next eight years saw the production of many important and valuable works, amongst which the "Geographical Distribution of Animals" (1876) occupies the chief place. This work, though perhaps the least known to the average reader, was considered by Wallace to be the most important scientific work he ever attempted. From references in letters written during his stay in the Malay Archipelago, it is clear that the subject had a strong attraction for him, and formed a special branch of study and observation many years before he began to work it out systematically in writing. His decision to write the book was the outcome of a suggestion made to him by Prof. A. Newton and Dr. Sclater about 1872. In addition to having already expressed his general views on this subject in various papers and articles, he had, after careful consideration, come to adopt Dr. Sclater's division of the earth's surface into six great zoological regions, which he found equally applicable to birds, mammalia, reptiles, and other great divisions; while at the same time it helped to explain the apparent contradictions in the distribution of land animals. Some years later he wrote:

In whatever work I have done I have always aimed at systematic arrangement and uniformity of treatment throughout. But here the immense extent of the subject, the overwhelming mass of detail, and above all the excessive diversities in the amount of knowledge of the different classes of animals, rendered it quite impossible to treat all alike. My preliminary studies had already satisfied me that it was quite useless to attempt to found any conclusions on those groups which were comparatively little known, either as regards the proportion of species collected and described, or as regards their systematic classification. It was also clear that as the present distribution of animals is necessarily due to their past distribution, the greatest importance must be given to those groups whose fossil remains in the more recent strata are the most abundant and the best known. These considerations led me to limit my work in its detailed systematic groundwork, and study of the principles and law of distribution, to the mammalia and birds, and to apply the principles thus arrived at to an explanation of the distribution of other groups, such as reptiles, fresh-water fishes, land and fresh-water shells, and the best-known insect Orders.

There remained another fundamental point to consider. Geographical distribution in its practical applications and interest, both to students and to the general reader, consists of two distinct divisions, or rather, perhaps, may be looked at from two points of view. In the first of these we divide the earth into regions and sub-regions, study the causes which have led to the difference in their animal productions, give a general account of these, with the amount of resemblance to and difference from other regions; and we may also give lists of the families and genera inhabiting each, with indications as to which are peculiar and which are also found in adjacent regions. This aspect of the study I term zoological geography, and it is that which would be of most interest to the resident or travelling naturalist, as it would give him, in the most direct and compact form, an indication of the numbers and kinds of animals he might expect to meet with.[1]

The keynote of the general scheme of distribution, as set forth in these two volumes, may be expressed as an endeavour to compare the extinct and existing fauna of each country and to trace the course by which what is now peculiar to each region had come to assume its present character. The main result being that all the higher forms of life seem to have originally appeared in the northern hemisphere, which has sent out migration after migration to colonise the three southern continents; and although varying considerably from time to time in form and extent, each has kept essentially distinct, while at the same time receiving periodically wave after wave of fresh animal life from the northward.

This again was due to many physical causes such as peninsulas parting from continents as islands, islands joining and making new continents, continents breaking up or effecting junction with or being isolated from one another. Thus Australia received the germ of her present abundant fauna of pouched mammals when she was part of the Old-World continent, but separated from that too soon to receive the various placental mammals which have, except in her isolated area, superseded those older forms. So, also, South America, at one time unconnected with North America, developed her great sloths and armadilloes, and, on fusing with the latter, sent her megatheriums to the north, and received mastodons and large cats in exchange.

Some of the points, such for instance as the division of the sub-regions into which each greater division is separated, gave rise to considerable controversy. Wallace's final estimate of the work stands: "No one is more aware than myself of the defects of the work, a considerable portion of which are due to the fact that it was written a quarter of a century too soon—at a time when both zoological and palæontological discovery were advancing with great rapidity, while new and improved classifications of some of the great classes and orders were in constant progress. But though many of the details given in these volumes would now require alteration, there is no reason to believe that the great features of the work and general principles established by it will require any important modification."[4]

About this time he wrote the article on "Acclimatisation" for the "Encyclopædia Britannica"; and another on "Distribution-Zoology" for the same work. As President of the Biological Section of the British Association he prepared an address for the meeting at Glasgow; wrote a number of articles and reviews, as well as his remarkable book on "Miracles and Modern Spiritualism." In 1878 he published "Tropical Nature," in which he gave a general sketch of the climate, vegetation, and animal life of the equatorial zone of the tropics from his own observations in both hemispheres. The chief novelty was, according to his own opinion, in the chapter on "climate," in which he endeavoured to show the exact causes which produce the difference between the uniform climate of the equatorial zone, and that of June and July in England. Although at that time *we* receive actually more of the light and heat of the sun than does Java or Trinidad in December, yet these places have then a mean temperature very much higher than ours. It contained also a chapter on humming-birds, as illustrating the luxuriance of tropical nature; and others on the colours of animals and of plants, and on various biological problems.[5]

"Island Life"[6] (published 1880) was begun in 1877, and occupied the greater part of the next three years. This had been suggested by certain necessary limitations in the writing of "The Geographical Distribution of Animals." It is a fascinating account of the relations of islands to continents, of their unwritten records of the distribution of plant and animal life in the morning time of the earth, of the causes and results of the glacial period, and of the manner of reckoning the age of the world from geological data. It also included several new features of natural science, and still retains an important place in scientific literature. No better summary can be given than that by the author himself:

In my "Geographical Distribution of Animals" I had, in the first place, dealt with the larger groups, coming down to families and genera, but taking no account of the various problems raised by the distribution of particular *species*. In the next place, I had taken little account of the various islands of the globe, excepting as forming sub-regions or parts of sub-regions. But I had long seen the great interest and importance of these, and especially of Darwin's great discovery of the two classes into which they are naturally divided—oceanic and continental islands. I had already given lectures on this subject, and had become aware of the great interest attaching to them, and the great light they threw upon the means of dispersal of animals and plants, as well as upon the past changes, both physical and means of dispersal and colonisation of animals is so connected with, and often dependent on, that of plants, that a consideration of the latter is essential to any broad views as to the distribution of life upon the earth, while they throw unexpected light upon those exceptional means of dispersal which, because they are exceptional, are often of paramount importance in leading to the production of new species and in thus determining the nature of insular floras and faunas.

Having no knowledge of scientific botany, it needed some courage, or, as some may think, presumption, to deal with this aspect of the problem; but ... I had long been excessively fond of plants, and ... interested in their distribution. The subject, too, was easier to deal with, on account of the much more complete knowledge of the detailed distribution of plants than of animals, and also because their classification was in a more advanced and stable condition. Again, some of the most interesting islands of the globe had been carefully studied botanically by such eminent botanists as Sir Joseph Hooker for the Galapagos, New Zealand, Tasmania, and the Antarctic islands; Mr. H.C. Watson for the Azores; Mr. J.G. Baker for Mauritius and other Mascarene islands; while there were floras by competent botanists of the Sandwich Islands, Bermuda and St. Helena....

But I also found it necessary to deal with a totally distinct branch of science— recent changes of climate as dependent on changes of the earth's surface, including the causes and effects of the glacial epoch, since these were among the most powerful agents in causing the dispersal of all kinds of organisms, and thus bringing about the actual distribution that now prevails. This led me to a careful study of Mr. James Croll's remarkable works on the subject of the astronomical causes of the glacial and interglacial periods.... While differing on certain details, I adopted the main features of his theory, combining with it the effects of changes in height and extent of land which form an important adjunct to the meteorological agents....

Besides this partially new theory of the causes of glacial epochs, the work contained a fuller statement of the various kinds of evidence proving that the great oceanic basins are permanent features of the earth's surface, than had before been given; also a discussion of the mode of estimating the duration of geological periods, and some considerations leading to the conclusion that organic change is now less rapid than the average, and therefore that less time is required for this change than has hitherto been thought necessary. I was also, I believe, the first to point out the great difference between the more ancient continental islands and those of more recent origin, with the interesting conclusions as to geographical changes afforded by both; while the most important novelty is the theory by which I explained the occurrence of northern groups of plants in all parts of the southern hemisphere—a phenomenon which Sir Joseph Hooker had pointed out, but had then no means of explaining.[7]

In 1878 Wallace wrote a volume on Australasia for Stanford's "Compendium of Geography and Travel." A later edition was published in 1893, which contained in addition to the physical geography, natural history, and geology of Australia, a much fuller account of the natives of Australia, showing that they are really a primitive type of the great Caucasian family of mankind, and are by no means so low in intellect as had been usually believed. This view has since been widely accepted.

Having, towards the close of 1885, received an invitation from the Lowell Institute, Boston, U.S.A., to deliver a course of lectures in the autumn and winter of 1886, Wallace decided upon a series which would embody those theories of evolution with which he was most familiar, with a special one on "The Darwinian Theory" illustrated by a set of original diagrams on variation. These lectures eventually became merged into the well-known book entitled "Darwinism."

On the first delivery of his lecture on the "Darwinian Theory" at Boston it was no small pleasure to Wallace to find the audience both large and attentive. One of the newspapers expressed the public appreciation in the following truly American fashion: "The first Darwinian, Wallace, did not leave a leg for anti-Darwinism to stand on when he had got through his first Lowell Lecture last evening. It was a masterpiece of condensed statement—as clear and simple as compact—a most beautiful specimen of scientific work. Dr. Wallace, though not an orator, is likely to become a favourite as a lecturer, his manner is so genuinely modest and straightforward."

Wherever he went during his tour of the States this lecture more than all others attracted and pleased his audiences. Many who had the opportunity of conversing with him, and others by correspondence, confessed that they had not been able to understand the "Origin of Species" until they heard the facts explained in such a lucid manner by him. It was this fact, therefore, which led him, on his return home in the autumn of 1887, to begin the preparation of the book ("Darwinism") published in 1889. The method he chose was that of following as closely as possible the lines of thought running through the "Origin of Species," to which he added many new features, in addition to laying special emphasis on the parts which had been most generally misunderstood. Indeed, so fairly and impartially did he set forth the general principles of the Darwinian theory that he was able to say: "Some of my critics declare that I am more Darwinian than Darwin himself, and in this, I admit, they are not far wrong."

His one object, as set out in the Preface, was to treat the problem of the origin of species from the standpoint reached after nearly thirty years of discussion, with an abundance of new facts and the advocacy of many new and old theories. As it had frequently been considered a weakness on Darwin's part that he based his evidence primarily on experiments with domesticated animals and cultivated plants, Wallace desired to secure a firm foundation for the theory in the variation of organisms in a state of nature. It was in order to make these facts intelligible that he introduced a number of diagrams, just as Darwin was accustomed to appeal to the facts of variation among dogs and pigeons.

Another change which he considered important was that of taking the struggle for existence first, because this is the fundamental phenomenon on which Natural Selection depends. This, too, had a further advantage in that, after discussing variations and the effects of artificial selection, it was possible at once to explain how Natural Selection acts.

The subjects treated with novelty and interest in their important bearings on the theory of Natural Selection were: (1) A proof that all *specific* characters are (or once have been) either useful in themselves or correlated with useful characters (Chap. VI.); (2) a proof that Natural Selection can, in certain cases, increase the sterility of crosses (Chap. VII.); (3) a fuller discussion of the colour relations of animals, with additional facts and arguments on the origin of sexual differences of colour (Chaps. VIII.-X.); (4) an attempted solution of the difficulty presented by the occurrence of both very simple and complex modes of securing the cross-fertilisation of plants (Chap. XI.); (5) some fresh facts and arguments on the wind-carriage of seeds, and its bearing on the wide dispersal of many arctic and alpine plants (Chap. XII.); (6) some new illustrations of the non-heredity of acquired characters, and a proof that the effects of use and disuse, even if inherited, must be overpowered by Natural Selection (Chap. XIV.); and (7) a new argument as to the nature and origin of the moral and intellectual faculties of man (Chap. XV.).

"Although I maintain, and even enforce," wrote Wallace, "my differences from some of Darwin's views, my whole work tends forcibly to illustrate the overwhelming importance of Natural Selection over all other agencies in the production of new species. I thus take up Darwin's earlier position, from which he somewhat receded in the later editions of his works, on account of criticisms and objections which I have endeavoured to show are unsound. Even in rejecting that phase of sexual selection depending on female choice, I insist on the greater efficacy of Natural Selection. This is pre-eminently the Darwinian doctrine, and I therefore claim for my book the position of being the advocate of pure Darwinism."

In concluding this section which, like a previous one, touches upon the intimate relations between Darwin and Wallace, and the points on which they agreed or differed, it is well, as the differences have been exaggerated and misunderstood, to bear in mind his own declaration: "None of my differences from Darwin imply any real divergence as to the overwhelming importance of the great principle of natural selection, while in several directions I believe that I have extended and strengthened it."[8]

With these explanatory notes the reader will now be able to follow the two groups of letters on Natural Selection, Geographical Distribution, and the Origin of Life and Consciousness which follow.

# II.—Correspondence on Biology, Geographical Distribution, etc.
## [1864-93]

### H. SPENCER TO A.R. WALLACE

*29 Bloomsbury Square, W.C. May 19, 1864.*

My dear Sir,—When I thanked you for your little pamphlet[2] the other day, I had not read it. I have since done so with great interest. Its leading idea is, I think, undoubtedly true, and of much importance towards an interpretation of the facts. Though I think that there are some purely physical modifications that may be shown to result from the direct influence of civilisation, yet I think it is quite clear, as you point out, that the small amounts of physical differences that have arisen between the various human races are due to the way in which mental modifications have served in place of physical ones.

I hope you will pursue the inquiry. It is one in which I have a direct interest, since I hope, hereafter, to make use of its results.—Sincerely yours,

HERBERT SPENCER

### SIR C. LYELL TO A.R. WALLACE

*53 Harley Street. May 22, [1864].*

My dear Sir,—I have been reading with great interest your paper on the Origin of the Races of Man, in which I think the question between the two opposite parties is put with such admirable clearness and fairness that that alone is no small assistance towards clearing the way to a true theory. The manner in which you have given Darwin the whole credit of the theory of Natural Selection is very handsome, but if anyone else had done it without allusion to your papers it would have been wrong.... With many thanks for your most admirable paper, believe me, my dear Sir, ever very truly yours,

CHA. LYELL.

### SIR C. LYELL TO A.R. WALLACE

*73 Harley Street. March 19, 1867.*

Dear Mr. Wallace,—I am citing your two papers in my second volume of the new edition of the "Principles"—that on the Physical Geography of the Malay Archipelago, 1863, and the other on Varieties of Man in ditto, 1864. I am somewhat confounded with the marked line which you draw between the two provinces on each side of the Straits of Lombok. It seems to me that Darwin and Hooker have scarcely given sufficient weight to the objection which it affords to some of their arguments. First, in regard to continental extension, if these straits could form such a barrier, it would seem as if nothing short of a land communication could do much towards fusing together two distinct faunas and floras. But here comes the question—are there any land-quadrupeds in Bali or in Lombok? I think you told me little was known of the plants, but perhaps you know something of the insects. It is impossible that birds of long flight crossing over should not have conveyed the seeds and eggs of some plants, insects, mollusca, etc. Then the currents would not be idle, and during such an eruption as that of Tomboro in Sumbawa all sorts of disturbances, aerial, aquatic and terrestrial, would have scattered animals and plants.

When I first wrote, thirty-five years ago, I attached great importance to preoccupancy, and fancied that a body of indigenous plants already fitted for every available station would prevent an invader, especially from, a quite foreign province, from having a chance of making good his settlement in a new country. But Darwin and Hooker contend that continental species which have been improved by a keen and wide competition are most frequently victorious over an insular or more limited flora and fauna. Looking, therefore, upon Bali as an outpost of the great Old World fauna, it ought to beat Lombok, which only represents a less rich and extensive fauna, namely the Australian.

You may perhaps answer that Lombok is an outpost of an army that may once have been as multitudinous as that of the old continent, but the larger part of the host have been swamped in the Pacific. But they say that European forms of animals and plants run wild in Australia and New Zealand, whereas few of the latter can do the same in Europe. In my map there is a small island called Nousabali; this ought to make the means of migration of seeds and animals less difficult. I cannot find that you say anywhere what is the depth of the sea between the Straits of Lombok, but you mention that it exceeds 100 fathoms. I am quite willing to infer that there is a connection between these soundings and the line of demarcation between the two zoological provinces, but must we suppose land communication for all birds of short flight? Must we unite South America with the Galapagos Islands? Can you refer me to any papers by yourself which might enlighten me and perhaps answer some of these queries? I should have thought that the intercourse even of savage tribes for tens of thousands of years between neighbouring islands would have helped to convey in canoes many animals and plants from one province to another so as to help to confound them. Your hypothesis of the gradual advance of two widely separated continents towards each other seems to be the best that can be offered. You say that a rise of a hundred fathoms would unite the Philippine Islands and Bali to the Indian region. Is there, then, a depth of 600 feet in that narrow strait of Bali, which seems in my map only two miles or so in breadth?

I have [been] confined to the house for a week by a cold or I should have tried to see you. I am afraid to go out to-day.—Believe me ever most truly yours,

CHA. LYELL.

## SIR C. LYELL TO A.R. WALLACE

*73 Harley Street. April 4, 1867.*

My dear Mr. Wallace,—I have been reading over again your paper published in 1855 in the *Annals* on "The Law which has regulated the Introduction of New Species"; passages of which I intend to quote, not in reference to your priority of publication, but simply because there are some points laid down more clearly than I can find in the work of Darwin itself, in regard to the bearing of the geological and zoological evidence on geographical distribution and the origin of species. I have been looking into Darwin's historical sketch thinking to find some allusion to your essay at page xx., 4th ed., when he gets to 1855, but I can find no allusion to it. Yet surely I remember somewhere a passage in which Darwin says in print that you had told him that in 1855 you meant by such expressions as "species being created on the type of pre-existing ones closely allied," and by what you say of modified prototypes, and by the passage in which you ask "what rudimentary organs mean if each species has been created independently," etc., that new species were created by variation and in the way of ordinary generation.

Your last letter was a great help to me, for it was a relief to find that the Lombok barrier was not so complete as to be a source of difficulty. I have also to thank you for your papers, one of which I had read before in the *Natural History Review*, but I am very glad of a separate copy. I am rather perplexed by Darwin speculating on the possibility of New Zealand having once been united with Australia (p. 446, 4th Ed.). The puzzle is greater than I can get over, even looking upon it as an oceanic island. Why should there have been no mammalia, rodents and marsupials, or only one mouse? Even if the Glacial period was such that it was enveloped in a Greenlandic winding-sheet, there would have been some Antarctic animals? It cannot be modern, seeing the height of those alps. It may have been a set of separate smaller islands, an archipelago since united into fewer. No savages could have extirpated mammalia, besides we should have found them fossil in the same places with all those species of extinct Dinornis which have come to light. Perhaps you will say that the absence of mammalia in New Caledonia is a corresponding fact.

This reminds me of another difficulty. On the hypothesis of the coral islands being the last remnants of a submerged continent, ought they not to have in them a crowd of peculiar and endemic types, each rivalling St. Helena, instead of which I believe they are very poor [in] peculiar genera. Have they all got submerged for a short time during the ups and downs to which they have been subjected, Tahiti and some others having been built up by volcanic action in the Pliocene period? Madeira and the Canaries were islands in the Upper Miocene ocean, and may therefore well have peculiar endemic types of very old date, and destroyed elsewhere. I have just got in Wollaston's "Coleoptera Atlantidum," and shall be glad to lend it you when I have read the Introduction. He goes in for continental extension, which only costs him two catastrophes by which the union and disunion with the nearest mainland may readily be accomplished.... —Believe me ever most truly yours,

CHA. LYELL.

## SIR C. LYELL TO A.R. WALLACE

*73 Harley Street. May 2, 1867.*

My dear Sir,—I forgot to ask you last night about an ornithological point which I have been discussing with the Duke of Argyll. In Chapter V. of his "Reign of Law" (which I should be happy to lend you, if you have time to look at it immediately) he treats of humming-birds, saying that Gould has made out about 400 species, every one of them very distinct from the other, and only one instance, in Ecuadór, of a species which varies in its tail-feathers in such a way as to make it doubtful whether it ought to rank as a species, an opinion to which Gould inclines, or only as a variety or incipient species, as the Duke thinks. For the Duke is willing to go so far towards the transmutation theory as to allow that different humming-birds may have had a common ancestral stock, provided it be admitted that a new and marked variety appears at once with the full distinctness of sex so remarkable in that genus.

According to his notion, the new male variety and the female must both appear at once, and this new race or species must be regarded as an "extraordinary birth." My reason for troubling you is merely to learn, since you have studied the birds of South America, and I hope collected some humming-birds, whether Gould is right in saying that there are so many hundred very distinct species without instances of marked varieties and transitional forms. If this be the case, would it not present us with an exception to the rule laid down by Darwin and Hooker that when a genus is largely represented in a continuous tract of land the species of that genus tend to vary?

I have inquired of Sclater and he tells me that he has a considerable distrust of Gould's information on this point, but that he has not himself studied humming-birds.

In regard to shells, I have always found that dealers have a positive prejudice against intermediate forms, and one of the most philosophical of them, now no more, once confessed to me that it was very much against his trade interest to give an honest opinion that certain varieties were not real species, or that certain forms, made distinct genera by some conchologists, ought not so to rank. Nine-tenths of his customers, if told that it was not a good genus or good species, would say, "Then I need not buy it." What they wanted was names, not things. Of course there are genera in which the species are much better defined than in others, but you would explain this, as Darwin and Hooker do, by the greater length of time during which they have existed, or the greater activity of changes, organic and inorganic, which have taken place in the region inhabited by the generic or family type in question. The manufactory of new species has ceased, or nearly so, and in that case I suppose a variety is more likely to be one of the transitional links which has not yet been extinguished than the first step towards a new permanent race or allied species....

Your last letter will be of great use to me. I had cited the case of beetles recovering from immersion of hours in alcohol from my own experience, but am glad it strikes you in the same light. McAndrew told me last night that the littoral shells of the Azores being European, or rather African, is in favour of a former continental extension, but I suspect that the floating of seaweed containing their eggs may dispense with the hypothesis of the submersion of 1,200 miles of land once intervening. I want naturalists carefully to examine floating seaweed and pumice met with at sea. Tell your correspondents to look out. There should be a microscopic examination of both these means of transport.—Believe me ever truly yours,

CHA. LYELL.

## SIR C. LYELL TO A.R. WALLACE

*73 Harley Street. July 3, 1867.*

My dear Mr. Wallace,—I was very glad, though I take in the *Westminster Review*, to have a duplicate of your most entertaining and instructive essay on Mimicry of Colours, etc., which I have been reading with great delight, and I may say that both copies are in full use here. I think it is admirably written and most persuasive.—Believe me ever most truly yours,

CHA. LYELL.

## TO HERBERT SPENCER

*Hurstpierpoint, Sussex. October 26, 1867.*

My dear Mr. Spencer,—After leaving you yesterday I thought a little over your objections to the Duke of Argyll's theory of flight on the ground that it does not apply to insects, and it seems to me that exactly the same general principles do apply to insects as to birds. I read over the Duke's book without paying special attention to that part of it, but as far as I remember, the case of insects offers no difficulty in the way of applying his principles. If any wing were a rigid plane surface, it appears to me that there are only two ways in which it could be made to produce flight. Firstly, on the principle that the resistance in a fluid, and I believe also in air, increases in a greater ratio than the velocity (? as the square), the descending stroke might be more rapid than the ascending one, and the resultant would be an upward or forward motion. Secondly, some kind of furling or feathering by a rotatory motion of the wing might take place on raising the wings. I think, however, it is clear that neither of these actions occurs during the flight of insects. In both slow- and quick-flying species there is no appearance of such a difference of velocity, and I am not aware that anyone has attempted to prove that it occurs; and the fact that in so many insects the edges of the fore and hind wings are connected together, while their insertions at the base are at some distance apart, *entirely precludes a rotation of the wings*. The whole structure and form of the wings of insects, moreover, indicate an action in flight quite analogous to that of birds. I believe that a careful examination will show that the wings of almost all insects are slightly concave beneath. Further, they are all constructed with a strong and rigid anterior margin, while the outer and hinder margins are exceedingly thin and flexible. Yet further, I feel confident (and a friend here agrees with me) that they are much more rigid against *upward* than against *downward* pressure. Now in most insects (take a butterfly as an example) the body is weighted behind the insertion of the wings by the long and heavy abdomen, so as to produce an oblique position when freely suspended. There is also much more wing surface behind than before the fulcrum. Now if such an insect produces by muscular action a regular flapping of the wings, flight must result. At the downward stroke the pressure of the air against the hind wings would raise them all to a nearly horizontal position, and at the same time bend up their posterior margins a little, producing an upward and onward motion. At the upward stroke the pressure on the hind wings would depress them considerably into an oblique position, and from their great flexibility in that direction would bend down their hind margins. The resultant would be a slightly downward and considerably onward motion, the two strokes producing that undulating flight so characteristic of butterflies, and so especially observable in the broad-winged tropical species. Now all this is quite conformable to the action of a bird's wing. The rigid anterior margin, the slender and flexible hind margin; the greater resistance to upward than to downward pressure, and the slight concavity of the under surface, are all characters common to the wings of birds and most insects, and, considering the totally different structure and homologies of the two, I think there is at least an *a priori* case for the function they both subserve being dependent upon these peculiarities. If I remember rightly, it is on these principles that the Duke of Argyll has explained the flight of birds, in which, however, there are of course some specialities depending on the more perfect organisation of the wing, its greater mobility and flexibility, its capacity for enlargement and contraction, and the peculiar construction and arrangement of the feathers. These, however, are matters of detail; and there are no doubt many and important differences of detail in the mode of flight of the different types of insects which would require a special study of each. It appeared to me that the Duke of Argyll had given that special study to the flight of birds, and deserved praise for having done so successfully, although he may not have quite solved the whole problem, or have stated quite accurately the comparative importance of the various causes that combine to effect flight.

—Believe me yours very sincerely,

ALFRED R. WALLACE.

## HERBERT SPENCER TO A.R. WALLACE

*57 Queen's Gardens, Bayswater, W. December 5, 1867.*

My dear Mr. Wallace,—I did not answer your last letter, being busy in getting out my second edition of "First Principles."

I was quite aware of the alleged additional cause of flight which you name, and do not doubt that it is an aid. But I regard it simply as an aid. If you will move an outstretched wing backwards and forwards with equal velocity, I think you will find that the difference of resistance is nothing like commensurate with the difference in size between the muscles that raise the wings and the muscles that depress them. It seems to me quite out of the question that the principles of flight are fundamentally different in a bat and a bird, which they must be if the Duke of Argyll's interpretation is correct. I write, however, not so much to reply to your argument as to correct a misapprehension which my expressions seem to have given you. The objections are not made by Tyndall or Huxley; but they are objections made by me, which I stated to them, and in which they agreed—Tyndall expressing the opinion that I ought to make them public. I name this because you may otherwise some day startle Tyndall or Huxley by speaking to them of *their* objections, and giving me as the authority for so affiliating them.—Very truly yours,

HERBERT SPENCER.

## SIR C. LYELL TO A.R. WALLACE

*73 Harley Street, London, W. November, 1867.*

Dear Wallace,—You probably remember an article by Agassiz in an American periodical, the *Christian Observer*, on the diversity of human races, etc., to prove that each distinct race was originally created for each zoological and botanical province. But while he makes out a good case for the circumscription of the principal races to distinct provinces, he evades in a singular manner the community of the Red Indian race to North and South America. He takes pains to show that the same American race pervades North and South America, or at least all America south of the Arctic region. This was Dr. Morton's opinion, and is, I suppose, not to be gainsaid. In other words, while the Papuan, Indo-Malayan, Negro and other races are strictly limited each of them to a particular region of mammalia, the Red Indian type is common to Sclater's Neo-arctic and Neo-tropical regions. Have you ever considered the explanation of this fact on Darwinian principles? If there were not barbarous tribes like the Fuegians, one might imagine America to have been peopled when mankind was somewhat more advanced and more capable of diffusing itself over an entire continent. But I cannot well understand why isolation such as accompanies a very low state of social progress did not cause the Neo-tropical and Neo-arctic regions to produce by varieties and Natural Selection two very different human races. May it be owing to the smaller lapse of time, which time, nevertheless, was sufficient to allow of the spread of the representatives of one and the same type from Canada to Cape Horn? Have you ever touched on this subject, or can you refer me to anyone who has?—Believe me ever most truly yours,

CHA. LYELL.

## TO SIR C. LYELL

1867.

Dear Sir Charles,—Why the colour of man is sometimes constant over large areas while in other cases it varies, we cannot certainly tell; but we may well suppose it to be due to its being more or less correlated with constitutional characters favourable to life. By far the most common colour of man is a warm brown, not very different from that of the American Indian. White and black are alike deviations from this, and are probably correlated with mental and physical peculiarities which have been favourable to the increase and maintenance of the particular race. I shall infer, therefore, that the brown or red was the original colour of man, and that it maintains itself throughout all climates in America because accidental deviations from it have not been accompanied by any useful constitutional peculiarities. It is Bates's opinion that the Indians are recent immigrants into the tropical plains of South America, and are not yet fully acclimatised.—Yours faithfully,

A.R. WALLACE.

## SIR C. LYELL TO A.R. WALLACE

*73 Harley Street. March 13, 1869.*

Dear Wallace,— ...I am reading your new book,[10] of which you kindly sent me a copy, with very great pleasure. Nothing equal to it has come out since Darwin's "Voyage of the *Beagle*." ... The history of the Mias is very well done. I am not yet through the first volume, but my wife is deep in the second and much taken with it. It is so rare to be able to depend on the scientific knowledge and accuracy of those who have so much of the wonderful to relate....—Believe me ever most truly yours,

CHA. LYELL.

## CANON KINGSLEY TO A.R. WALLACE

*Eversley Rectory, Winchfield. May 5, 1869.*

My dear Sir,—I am reading—or rather have all but read—your new book,[11] with a delight which I cannot find words to express save those which are commonplace superlatives. Let me felicitate you on having, at last, added to the knowledge of our planet a chapter which has not its equal (as far as I can recollect) since our friend Darwin's "Voyage of the *Beagle*." Let me, too, compliment you on the modesty and generosity which you have shown, in dedicating your book to Darwin, and speaking of him and his work as you have done. Would that a like unselfish chivalry were more common—I do not say amongst scientific men, for they have it in great abundance, but—in the rest of the community.

May I ask—as a very great favour—to be allowed to call on you some day in London, and to see your insects? I and my daughter are soon, I hope, going to the West Indies, for plants and insects, among other things; and the young lady might learn much of typical forms from one glance at your treasures.

I send this letter by our friend Bates—being ignorant of your address.—Believe me, my dear Sir, ever yours faithfully,

C. KINGSLEY.

### TO MISS A. BUCKLEY[12]

*Holly House, Barking, E. February 2, 1871.*

Dear Miss Buckley,—I have read Darwin's first volume,[13] and like it very much. It is overwhelming as proving the origin of man from some lower form, but that, I rather think, hardly anyone doubts now.

He is very weak, as yet, on my objection about the "hair," but promises a better solution in the second volume.

Have you seen Mivart's book, "Genesis of Species"? It is exceedingly clever, and well worth reading. The arguments against Natural Selection as the exclusive mode of development are some of them exceedingly strong, and very well put, and it is altogether a most readable and interesting book.

Though he uses some weak and bad arguments, and underrates the power of Natural Selection, yet I think I agree with his conclusion in the main, and am inclined to think it is more philosophical than my own. It is a book that I think will please Sir Charles Lyell.—Believe me, yours very truly,

ALFRED R. WALLACE.

### TO MISS A. BUCKLEY

*Holly House, Barking, E. March 3, 1871.*

Dear Miss Buckley,—Thanks for your note. I am hard at work criticising Darwin. I admire his Moral Sense chapter as much as anything in the book. It is both original and the most satisfactory of all the theories, if not quite satisfactory....—Believe me yours very faithfully,

ALFRED R. WALLACE.

P.S.—Darwin's book on the whole is wonderful! There are plenty of points open to criticism, but it is a marvellous contribution to the history of the development of the forms of life.

## SIR C. LYELL TO A.R. WALLACE

*February 15, 1876.*

Dear Wallace,—I have read the Preface,[14] and like and approve of it much. I do not believe there is a word which Darwin would wish altered. It is high time this modest assertion of your claims as an independent originator of Natural Selection should be published.—Ever most truly,

CHA. LYELL.

## SIR J. HOOKER TO A.R. WALLACE

*Royal Gardens, Kew. August 2, 1880.*

My dear Wallace,—I think you have made an immense advance to our knowledge of the ways and means of distribution, and bridged many great gaps.[15] Your reasoning seems to me to be sound throughout, though I am not prepared to receive it in all its details.

I am disposed to regard the Western Australian flora as the latest in point of origin, and I hope to prove it by development, and by the absence of various types. If Western Australia ever had an old flora, I am inclined to suppose that it has been destroyed by the invasion of Eastern types after the union with East Australia. My idea is that these types worked round by the south, and altered rapidly as they proceeded westward, increasing in species. Nor can I conceive the Western Island, when surrounded by sea, harbouring a flora like its present one.

I have been disposed to regard New Caledonia and the New Hebrides as the parent country of many New Zealand and Australian forms of vegetation, but we do not know enough of the vegetation of the former to warrant the conclusion; and after all it would be but a slight modification of your views.

I very much like your whole working of the problem of the isolation and connection of New Zealand and Australia *inter se* and with the countries north of them, and the whole treatment of that respecting north and south migration over the globe is admirable....—Ever most truly yours,

J.D. HOOKER.

## SIR J. HOOKER TO A.R. WALLACE

*Royal Gardens, Kew. November 10, 1880.*

Dear Mr. Wallace,—I have been waiting to thank you for "Island Life" till I should have read it through as carefully as I am digesting the chapters I have finished; but I can delay no longer, if only to say that I heartily enjoy it, and believe that you have brushed away more cobwebs that have obscured the subject than any other, besides giving a vast deal that is new, and admirably setting forth what is old, so as to throw new light on the whole subject. It is, in short, a first-rate book. I am making notes for you, but hitherto have seen no defect of importance except in the matter of the Bahamas, whose flora is Floridan, not Cuban, in so far as we know it....—Very truly yours,

JOS. D. HOOKER.


## TO SIR W. THISELTON-DYER

*Pen-y-bryn, St. Peter's Road, Croydon. January 7, 1881.*

Dear Mr. Thiselton-Dyer,—If I had had your lecture before me when writing the last chapters of my book I should certainly have quoted you in support of the view of the northern origin of the Southern flora by migration along existing continents. On reading it again I am surprised to find how often you refer to this; but when I read it on its first appearance I did not pay special attention to this point except to note that your views agreed more closely with those I had advanced, derived from the distribution of animals, than those of any previous writer on botanical distribution. When, at a much later period, on coming to the end of my work, I determined to give a chapter to the New Zealand flora in order to see how far the geological and physical relations between New Zealand and Australia would throw light on its origin, I went for my facts to the works of Sir Joseph Hooker and Mr. Bentham, and also to your article in the "Encyclopædia Britannica," and worked out my conclusions solely from these, and from the few facts referring to the migration of plants which I had collected. Had I referred again to your lecture I should certainly have quoted the cases you give (in a note, p. 431) of plants extending along the Andes from California to Peru and Chile, and vice versa. Whatever identity there is in our views was therefore arrived at independently, and it was an oversight on my part not referring to your views, partly due to your not having made them a more prominent feature of your very interesting and instructive lecture. Working as I do at home, I am obliged to get my facts from the few books I can get together; and I only attempted to deal with these great botanical questions because the facts seemed sufficiently broad and definite not to be much affected by errors of detail or recent additions to our knowledge, and because the view which I took of the past changes in Australia and New Zealand seemed calculated to throw so much light upon them. Without such splendid summaries of the relations of the Southern floras as are given in Sir J. Hooker's Introductions, I should not have touched the subject at all; and I venture to hope that you or some of your colleagues will give us other such summaries, brought down to the present date, of other important floras—as, for example, those of South Africa and South Temperate America.

Many thanks for additional peculiar British plants. When I hear what Mr. Mitten has to say about the mosses, etc., I should like to send a corrected list to *Nature*, which I shall ask you to be so good as to give a final look over.— Believe me yours very faithfully,

ALFRED R. WALLACE.

P.S.—Mr. Darwin strongly objects to my view of the migration of plants along mountain-ranges, rather than along lowlands during cold periods. This latter view seems to me as difficult and inadequate as mine does to him.— A.R.W.


Wallace was in frequent correspondence with Professor Raphael Meldola, the eminent chemist, a friend both of Darwin and of Wallace, a student of Evolution, and a stout defender of Darwinism. I received from him much help and advice in connection with this work, and had he lived until its completion—he died, suddenly, in 1914—my indebtedness to him would have been even greater.

The following letter to Meldola refers to a suggestion that the white colour of the undersides of animals might have been developed by selection through the *physical* advantage gained from the protection of the vital parts by a *lighter* colour and therefore by a surface of less radiative activity. The idea was that there would be less loss of animal heat through such a white coating. We were at that time unaware of Thayer's demonstration of the value of such colouring for the purposes of concealment among environment. Wallace accepted Thayer's view at once when it was subsequently put forward; as do most naturalists at the present time.

## TO PROF. MELDOLA

*Frith Hill, Godalming. April 8, 1885.*

My dear Meldola,—Your letter in *Nature* last week "riz my dander," as the Yankees say, and, for once in a way, we find ourselves deadly enemies prepared for mortal combat, armed with steel (pens) and prepared to shed any amount of our own—ink. Consequently I rushed into the fray with a letter to *Nature* intended to show that you are as wrong (as wicked) as are the Russians in Afghanistan. Having, however, the most perfect confidence that the battle will soon be over, ... —Yours very faithfully,

ALFRED R. WALLACE.


The following letter refers to the theory of physiological selection which had recently been propounded by Romanes, and which Prof. Meldola had criticised in *Nature*, xxxix. 384.


## TO PROF. MELDOLA

*Frith Hill, Godalming. August 28, 1886.*

My dear Meldola,—I have just read your reply to Romanes in *Nature*, and so far as your view goes I agree, but it does not go far enough. Professor Newton has called my attention to a passage in Belt's "Nicaragua," pp. 207-8, in which he puts forth very clearly exactly your view. I find I had noted the explanation as insufficient, and I hear that in Darwin's copy there is "No! No!" against it. It seems, however, to me to summarise *all* that is of the slightest value in Romanes' wordy paper. I have asked Newton (to whom I had lent it) to forward to you at Birmingham a proof of my paper in the *Fortnightly*, and I shall be much obliged if you will read it carefully, and, if you can, "hold a brief" for me at the British Association in this matter. You will see that a considerable part of my paper is devoted to a demonstration of the fallacy of that part of "Romanes" which declares species to be distinguished generally by useless characters, and also that "simultaneous variations" do not usually occur.

On the question of sterility, which, as you well observe, is the core of the question, I think I show that it could not work in the way Romanes puts it. The objection to Belt's and your view is, also, that it would not work unless the "sterility variation" was correlated with the "useful variation." You assume, I think, this correlation, when you speak of two of your varieties, B. and K., being *less fertile with the parent form*. Without correlation they could not be so, only some few of them. Romanes always speaks of his physiological variations as being independent, "primary," in which case, as I show, they could hardly ever survive. At the end of my paper I show a correlation which is probably general and sufficient.

In criticising Romanes, however, at the British Association, I want to call your special attention to a point I have hardly made clear enough in my paper. Romanes always speaks of the "physiological variety" as if it were like any other *simple* variety, and could as easily (he says more easily) be increased. Whereas it is really complex, requiring a remarkable correlation between different sets of individuals which he never recognises. To illustrate what I mean, let me suppose a case. Let there occur in a species three individual physiological varieties—A, B and C—each being infertile with the bulk of the species, but quite fertile with some small part of it. Let A, for example, be fertile with X, Y and Z. Now I maintain it to be in the highest degree improbable that B, a quite distinct individual, with distinct parents originating in a distinct locality, and perhaps with a very different constitution, merely because it also is sterile with the bulk of the species, should be fertile with the very same individuals, X, Y, Z, that A is fertile with. It seems to me to be at least 100 to 1 that it will be fertile with some other quite distinct set of individuals. And so with C, and any other similar variety. I express this by saying that each has its "sexual complements," and that the complements of the one are almost sure not to be the complements of the other. Hence it follows that A, B, C, though differing in the same character of general infertility with the bulk of the species, will really be three distinct varieties physiologically, and can in no way unite to form a single physiological variety. This enormous difficulty Romanes apparently never sees, but argues as if all individuals that are infertile with the bulk of the species must be or usually are fertile with the same set of individuals or with each other. This I call a monstrous assumption, for which not a particle of evidence exists. Take this in conjunction with my argument from the severity of the struggle for existence and the extreme improbability of the respective "sexual complements" coming together at the right time, and I think Romanes' ponderous paper is disposed of.

I wrote my paper, however, quite as much to expose the great presumption and ignorance of Romanes in declaring that Natural Selection is *not* a theory of the origin of species—as it is calculated to do much harm. See, for instance, the way the Duke of Argyll jumped at it like a trout at a fly!—Yours very faithfully,

ALFRED R. WALLACE.

The earlier part of the next letter refers to "The Experimental Proof of the Protective Value of Colour and Markings in Insects in reference to their

Vertebrate Enemies," in the *Proceedings of the Zoological Society of London*, 1887, p. 191.

## TO PROF. POULTON

*Frith Hill, Godalming. October 20, 1887.*

My dear Poulton,—It is very interesting to me to see how very generally the facts are in accordance with theory, and I am only surprised that the exceptions and irregularities are not more numerous than they are found to be. The only difficult case, that of *D. euphorbiæ*, is due probably to incomplete knowledge. Are lizards and sea-birds the only, or even the chief, possible enemies of the species? They evidently do not prevent its coming to maturity in considerable abundance, and it is therefore no doubt preserved from its chief enemies during its various stages of growth.

The only point on which I differ from you—as you know—is your acceptance, as proved, of the theory of sexual colour selection, and your speaking of insects as having a sense of "the beautiful" in colour, as if that were a known fact. But that is a wide question, requiring full discussion.—Yours very faithfully,

ALFRED R. WALLACE.

## TO SIR FRANCIS DARWIN

*Frith Hill, Godalming. November 20, 1887.*

Dear Mr. Darwin,—Many thanks for the copy of your father's "Life and Letters," which I shall read with very great interest (as will all the world). I was not aware before that your father had been so distressed—or rather disturbed—by my sending him my essay from Ternate, and I am very glad to feel that his exaggerated sense of honour was quite needless so far as I was concerned, and that the incident did not in any way disturb our friendly relations. I always felt, and feel still, that people generally give me far too much credit for my mere sketch of the theory—so very small an affair as compared with the vast foundation of fact and experiment on which your father worked.—Believe me yours very faithfully,

ALFRED R. WALLACE.

## TO MRS. FISHER (*née* BUCKLEY)

*Frith Hill, Godalming. February 16, 1888.*

My dear Mrs. Fisher,—I know nothing of the physiology of ferns and mosses, but as a matter of fact I think they will be found to increase and diminish together all over the world. Both like moist, equable climates and shade, and are therefore both so abundant in oceanic islands, and in the high regions of the tropics.

I am inclined to think that the reason ferns have persisted so long in competition with flowering plants is the fact that they thrive best in shade, flowers best in the light. In our woods and ravines the flowers are mostly spring flowers, which die away just as the foliage of the trees is coming out and the shade deepens; while ferns are often dormant at that time, but grow as the shade increases.

Why tree-ferns should not grow in cold countries I know not, except that it may be the winds are too violent and would tear all the fronds off before the spores were ripe. Everywhere they grow in ravines, or in forests where they are sheltered, even in the tropics. And they are not generally abundant, but grow in particular zones only. In all the Amazon valley I don't remember ever having seen a tree-fern....

I too am struggling with my "Popular Sketch of Darwinism," and am just now doing a chapter on the great "hybridity" question. I really think I shall be able to arrange the whole subject more intelligibly than Darwin did, and simplify it immensely by leaving out the endless discussion of collateral details and difficulties which in the "Origin of Species" confuse the main issue....

The most remarkable steps yet made in advance are, I think, the theory of Weismann of the continuity of the germ plasm, and its corollary that acquired modifications are never inherited! and Patrick Geddes's explanation of the laws of growth in plants on the theory of the antagonism of vegetative and reproductive growth....—Yours very sincerely,

ALFRED R. WALLACE.

## TO PROF. MELDOLA

*Frith Hill, Godalming. March 20, 1888.*

My dear Meldola,—I have been working away at my hybridity chapters,[16] and am almost disposed to cry "Eureka!" for I have got light on the problem. When almost in despair of making it clear that Natural Selection could act one way or the other, I luckily routed out an old paper that I wrote twenty years ago, giving a demonstration of the action of Natural Selection. It did not convince Darwin then, but it has convinced me now, and I think it can be proved that in some cases (and those I think most probable) Natural Selection will accumulate variations in infertility between incipient species. Many other causes of infertility co-operate, and I really think I have overcome the fundamental difficulties of the question and made it a good deal clearer than Darwin left it.... I think also it completely smashes up Romanes.—Yours faithfully,

ALFRED R. WALLACE.

The next letter relates to a question which Prof. Meldola raised as to whether, in view of the extreme importance of "divergence" (in the Darwinian sense) for the separation and maintenance of specific types, it might not be possible that sterility, when of advantage as a check to crossing, had in itself, as a physiological character, been brought about by Natural Selection, just as extreme fecundity had been brought about (by Natural Selection) in cases where such fecundity was of advantage.

## TO PROF. MELDOLA

*Frith Hill, Godalming. April 12, 1888.*

My dear Meldola,—Many thanks for your criticism. It is a perfectly sound one as against my view being a *complete explanation* of the phenomena, but that I do not claim. And I do not see any chance of the required facts being forthcoming for many years to come. Experiments in the hybridisation of animals are so difficult and tedious that even Darwin never undertook any, and the only people who could and ought to have done it—the Zoological Society—will not. There is one point, however, I think you have overlooked. You urge the improbability of the required infertility being correlated with the particular variations which characterised each incipient species. But the whole point of my argument is, that the physiological adjustments producing fertility are so delicate that they are disturbed by almost any variation or change of conditions—except in the case of domestic animals, which have been domesticated because they are not subject to this disturbance. The whole first half of the chapter is to bring out this fact, which Darwin has dwelt upon, and it certainly does afford a foundation for the assumption that usually, and in some considerable number of individuals, variation in nature, accompanied by somewhat changed conditions of life, is accompanied by, and probably correlated with, some amount of infertility. No doubt this assumption wants proving, but in the meantime I am glad you think that, granting the assumption, I have shown that Natural Selection is able to accumulate sterility variations.

That is certainly a step in advance, and we cannot expect to do more than take very short theoretical steps till we get more facts to rest upon. If you should happen to come across any facts which seem to bear upon it, pray let me know. I can find none but those I have referred to.

I have just finished a chapter on male ornament and display, which I trust will help to clear up that point—Believe me yours very faithfully,

ALFRED R. WALLACE.

## TO DR. W.B. HEMSLEY

*Frith Hill, Godalming. August 26, 1888.*

Dear Mr. Hemsley,—You are aware that Patrick Geddes proposes to exclude Natural Selection in the origination of thorns and spines, which he imputes to "diminishing vegetativeness" or "ebbing vitality of the species." It has occurred to me that insular floras should afford a test of the correctness of this view, since in the absence of mammalia the protection of spines would be less needed.

Your study of these floras will no doubt enable you to answer a few questions on this point. Spines and thorns are, I believe, usually abundant in arid regions of continents, especially in South Africa, where large herbivorous mammals abound. Now, if the long-continued presence of these mammals is a factor in the production of spines by Natural Selection, they should be wholly or comparatively absent in regions equally arid where there are no mammals. The Galapagos seem to be such a case—also perhaps some of the Sandwich Islands, and generally the extra-tropical volcanic islands. Also Australia comparatively, and the highlands of Madagascar.

Of course, the endemic species must be chiefly considered, as they have had time to be modified by the conditions. If you can give me the facts, or your general impression from your study of these floras, I shall be much obliged. I see, of course, many other objections to Geddes's theory, but this seems to offer a crucial test.—Believe me yours very truly,

ALFRED R. WALLACE.

## TO DR. W.B. HEMSLEY

*Frith Hill, Godalming. September 13, 1888.*

Dear Mr. Hemsley,—Many thanks for your interesting letter. The facts you state seem quite to support the usual view, that thorns and spines have been developed as a protection against other animals. The few spiny plants in New Zealand may be for protection against land molluscs, of which there are several species as large as any in the tropics. Of course in Australia we should expect only a comparative scarcity of spines, as there are many herbivorous marsupials in the country.—Believe me yours very faithfully,

ALFRED R. WALLACE.


The next and several of the succeeding letters refer to the translations of Weismann's "Essays upon Heredity and Kindred Biological Problems" (Oxford, 1889), and to "Darwinism" (London, 1889).

## TO PROF. POULTON

*Frith Hill, Godalming. November 4, 1888.*

My dear Mr. Poulton,—I returned you the two first of Weismann's essays, with a few notes and corrections in pencil on that on "Duration of Life." Looking over some old papers, I have just come across a short sketch on two pages, on "The Action of Natural Selection in producing Old Age, Decay and Death," written over twenty years ago.[17] I had the same general idea as Weismann, but not that beautiful suggestion of the duration of life, in each case, being the *minimum* necessary for the preservation of the species. *That* I think masterly. The paper on "Heredity" is intensely interesting, and I am waiting anxiously for the concluding part. I will refer to these papers in notes in my book, though perhaps yours will be out first....—Yours faithfully,

A.R. WALLACE.

## TO PROF. POULTON

*Frith Hill, Godalming. November 8, 1888.*

Dear Mr. Poulton,—I return herewith (but separately) the "proofs" I have of Weismann's Essays. The last critical one is rather heavy, and adds nothing of importance to the earlier one on Duration of Life. I enclose my "Note" on the subject, which was written, I think, about 1867, certainly before 1870. You will see it was only a few ideas jotted down for further elaboration and then forgotten. I see however it *does* contain the germ of Weismann's argument as to duration of life being determined by the time of securing continuance of the species.—Yours faithfully,

A.R. WALLACE.

## TO PROF. POULTON

*Frith Hall, Godalming. January 20, 1889.*

My dear Mr. Poulton,—My attention has been called by Mr. Herdman, in his Inaugural Address to the Liverpool Biological Society, to Galton's paper on "Heredity," which I read years ago but had forgotten. I have just read it again (in the *Journal of the Anthropological Institute*, Vol. V., p. 329, Jan., 1876), and I find a remarkable anticipation of Weismann's theories which I think should be noticed in a preface to the translation of his book.[18] He argues that it is the undeveloped germs or gemmules of the fertilised ovum that form the sexual elements of the offspring, and thus heredity and atavism are explained. He also argues that, as a corollary, "acquired modifications are barely if at all inherited in the correct sense of the word." He shows the imperfection of the evidence on this point, and admits, just as Weismann does, the heredity of changes in the parent like alcoholism, which, by permeating the whole tissues, may *directly* affect the reproductive elements. In fact, all the main features of Weismann's views seem to be here anticipated, and I think he ought to have the credit of it.

Being no physiologist, his language is not technical, and for this reason, and the place of publication perhaps, his remarkable paper appears to have been overlooked by physiologists.

I think you will find the paper very suggestive, even supplying some points overlooked by Weismann.—Yours faithfully,

A.R. WALLACE.

## TO PROF. POULTON

*Hamilton House, The Croft, Hastings. February 19, 1889.*

Dear Mr. Poulton,—Do you happen to have, or can you easily refer to, Grant Allen's small books of collected papers under such titles as "Vignettes from Nature," "The Evolutionist at Large," "Colin Clout's Calendar," and another I can't remember? In one of them is a paper on the Origin of Wheat, in which he puts forth the theory that the grasses, etc., are degraded forms which were once insect-fertilised, summing up his views in the phrase, "Wheat is a degraded lily," or something like that. Now Henslow, in his "Floral Structures,"[19] adopts the same theory for all the wind-fertilised or self-fertilised flowers, and he tells me that he is *alone* in the view. I believe the view is a true one, and I want to give G. Allen the credit of first starting it, and want to see how far he went. If you have or can get this work of his with that paper, can you lend it me for a few days? I know not who to write to for it, as botanists of course ignore it, and G. Allen himself is, I believe, in Algeria....—Yours faithfully,

ALFRED R. WALLACE.

## HERBERT SPENCER TO A.R. WALLACE

*38 Queen's Gardens, Lancaster Gate, W. May 18, 1889.*

Dear Mr. Wallace,—A few days ago there reached me a copy of your new book, "Darwinism," for which, along with this acknowledgment, I send my thanks. In my present state of health I dare not read, and fear I shall be unable to profit by the accumulation of evidence you have brought together. I see sundry points on which I might raise discussions, but beyond the fact that I am at present unable to enter into them, I doubt whether they would be of any use. I regret that you have used the title "Darwinism," for notwithstanding your qualification of its meaning you will, by using it, tend greatly to confirm the erroneous conception almost universally current.—Truly yours,

HERBERT SPENCER.

## TO PROF. POULTON

*Parkstone, Dorset. November 28, 1889.*

My dear Mr. Poulton,—I have much pleasure in sending you Cope's book[20] (with the review of "Darwinism"), which I hope you will keep as long as you like, till you have mastered all its obscurities of style and eccentricities of argument. I think you will find a good deal in it to criticise, and it will be well for you to know what the leader of the Neo-Lamarckians regards as the foundation-stones of his theory. I greatly enjoyed my visit to Oxford, and only regretted that I could not leave more time for personal talk with yourself, for I am so deplorably ignorant of modern physiology that I am delighted to get intelligible explanations of its bearings on the subjects that most interest me in science. I quite see all its importance in investigations of the mechanism of colours, but there is so much still unknown that it will be very hard to convince me that there is no other possible explanation of the peacock's feather than the "continued preference by the females" for the most beautiful males, in *this one point*, "during a long line of descent"—as Darwin says! I expect, however, great light from your new book....—Believe me yours very faithfully,

ALFRED R. WALLACE.

## SIR FRANCIS GALTON TO A.R. WALLACE

*42 Rutland Gate, S.W. May 24, 1890.*

Dear Mr. Wallace,—I send the paper with pleasure, and am glad that you will read it, and I hope then see more clearly than the abstract could show the grounds of my argument.

These finger-marks are most remarkable things. Of course I have made out much more about them since writing that memoir. Indeed I have another paper on them next Thursday at the Royal Society, but that only refers to ways of cataloguing them, either for criminal administration, or what I am more interested in, viz. racial and hereditary inquiry.

What I have done in this way is not ready for publication, but I may mention (privately, please) that these persistent marks, which seem fully developed in the sixth month of foetal life, and appear under the reservations and in the evidence published in the memoir to be practically *quite* unchanged during life, are *not* correlated with any ordinary characteristic that I can discover. They are the same in the lowest idiots as in ordinary persons. (I took the impressions of some 80 of these, so idiotic that they mostly could not speak, or even stand, at the great Darenth Asylum, Dartford.) They are the same in clod-hoppers as in the upper classes, and *yet* they are as hereditary as other qualities, I think. Their tendency to symmetrical distribution on the two hands is *marked*, and symmetry *is* a form of kinship. My argument is that sexual selection can have had nothing to do with the patterns, neither can any other form of selection due to vigour, wits, and so forth, because they are not correlated with them. They just go their own gait, uninfluenced by anything that we can find or reasonably believe in, of a *naturally selective influence*, in the plain meaning of the phrase.—Very sincerely yours,

FRANCIS GALTON.

## TO THEO. D.A. COCKERELL

*Parkstone, Dorset. March 10, 1891.*

Dear Mr. Cockerell,— ... Your theory to account for the influence of a first male on progeny by a second seems very probable—and in fact if, as I suppose, spermatozoa often enter ova without producing complete fertilisation, it must be so. *That* would be easily experimented on, with fowls, dogs, etc., but I do not remember the fact having been observed except with horses. It ought to be common, when females have young by successive males.—Yours faithfully,

A.R. WALLACE.

The next letter relates to a controversy with Romanes concerning Herbert Spencer's argument about Co-adaptation which Romanes had urged in support of Neo-Lamarckism as opposed to Natural Selection. Prof. Meldola endeavoured to show that the difficulties raised by Spencer and supported by Romanes had no real weight because the possibility of so-called "co-adaptations" being developed *successively* in the order of evolution had not been reckoned with. There was no real divergence between Wallace and Prof. Meldola on this matter when they subsequently discussed it. The correspondence is in *Nature*, xliii. 557, and subsequently. *See also* "Darwin and After Darwin," by Romanes, 1895, ii. 68.

## TO PROF. MELDOLA

*Parkstone, Dorset, April 25, 1891.*

My dear Meldola,—You have now put your foot in it! Romanes *agrees* with you! Henceforth he will claim you as a disciple, converted by his arguments!

There was one admission in your letter I was very sorry to see, because it cannot be strictly true, and is besides open to much misrepresentation. I mean the admission that Romanes pounces upon in his second paragraph. Of course, the number of individuals in a species being finite, the chance of four coincident variations occurring in any one individual—each such variation being separately very common—cannot be anything like "infinity to one." Why, then, do you concede it most fully?—the result being that Romanes takes you to concede that it is infinity to one against the coincident variations occurring in "*any individuals.*" Surely, with the facts of coincident independent variation we now possess, the occurrence of three, four, or five, coincident variations cannot be otherwise than frequent. As a fact, more than half the whole population of most species seems to vary to a perceptible and measurable, and therefore sufficient, amount in scores of ways. Take a species with a million pairs of individuals—half of these vary sufficiently, either + or -, in the four acquired characters A, B, C, D: what will be the proportion of individuals that vary + in these four characters according to the law of averages? Will it not be about 1 in 64? If so it is ample—in many cases—for Natural Selection to work on, because in many cases less than 1/64 of offspring survives.

On Romanes' view of the impossibility of Natural Selection doing anything alone, because the required coincident variations do not occur, the occurrence of a "strong man" or a racehorse that beats all others easily must be impossible, since in each of these cases there must be scores of coincident favourable variations.

Given sufficient variation, I believe divergent modification of a species in two lines could easily occur, even if free intercrossing occurred, because, the numbers varying being a large proportion of the whole, the numbers which bred like with like would he sufficient to carry on the two lines of divergence, those that intercrossed and produced less perfectly adapted offspring being eliminated. Of course some amount of segregate breeding does always occur, as Darwin always maintained, but, as he also maintained, it is not absolutely essential to evolution. Romanes argues as if "free intercrossing" meant that none would pair like with like! I hope you will have another slap at him, and withdraw or explain that unlucky "infinity to one," which is Romanes' sheet-anchor.—Yours very truly,

ALFRED R. WALLACE.

## TO PROF. POULTON

*Parkstone, Dorset. June 16, 1892.*

My dear Mr. Poulton,—Many thanks for sending me Weismann's additional Essays,[21] which I look forward to reading with much pleasure. I have, however, read the first, and am much disappointed with it. It seems to me the *weakest and most inconclusive* thing he has yet written. At p. 17 he states his theory as to degeneration of eyes, and again, on p. 18, of anthers and filaments; but in both cases he fails to *prove* it, and apparently does not see that his panmixia, or "cessation of selection," cannot possibly produce *continuous* degeneration culminating in the total or almost total disappearance of an organ. Romanes and others have pointed out this weakness in his theory, but he does not notice it, and goes on calmly throughout the essay to *assume* that mere panmixia must cause progressive degeneration to an unlimited extent; whereas all it can do is to effect a reduction to the average of the total population on which selection has been previously worked. He says "individuals with weak eyes would not be eliminated," but omits to notice that individuals with strong eyes would also "not be eliminated," and as there is no reason alleged why variations in *all directions* should not occur as before, the free intercrossing would tend to keep up a mean condition only a little below that which was kept up by selection. It is clear that some form of selection must always co-operate in degeneration, such as economy of growth, which he hardly notices except as a possible but not a necessary factor, or actual injuriousness. It appears to me that what is wanted is to take a number of typical cases, and in each of them show how Natural Selection comes in to carry on the degeneration begun by panmixia. Weismann's treatment of the subject is merely begging the question.—Yours faithfully,

A.R. WALLACE.

## TO PROF. POULTON

*Parkstone, Dorset. August 29, 1892.*

My dear Mr. Poulton,—As to panmixia you have quite misunderstood my position. By the "mean condition," I do not mean the "mean" during the whole course of development of the organ, as you seem to take it. That would indeed be absurd. I do mean the "mean" of the whole series of individual variations now occurring, during a period sufficient to contain all or almost all the variations to which the species is *now* subject. Take, for instance, such a case as the wings of the swallow, on the full development of which the life of the bird depends. Many individuals no doubt perish for lack of wing-power, due to deficiency in size or form of wing, or in the muscles which move it. The extreme limits of variation would be seen probably if we examined every swallow that had reached maturity during the last century. The average of all those would perhaps be 5 or 10 per cent. below the average of those that survive to become the parents of the next generation in any year; and what I maintain is, that panmixia alone could not reduce a swallow's wings below this first average. Any further reduction must be due either to some form of selection or to "economy of growth"—which is also, fundamentally, a form of selection. So with the eyes of cave animals, panmixia could only cause an imperfection of vision equal to the average of those variations which occurred, say, during a century before the animal entered the cave. It could only produce more effect than this if the effects of disuse are hereditary—which is a non-Weismannian doctrine. I think this is also the position that Romanes took.—Yours faithfully,

A.R. WALLACE.

## TO MR. J.W. MARSHALL

*Parkstone, Dorset. September 23, 1892.*

My dear Marshall,—I am glad you enjoyed Mr. Hudson's book. His observations are inimitable—and his theories and suggestions, if not always the best, at least show thought on what he has observed.

I was most pleased with his demonstration as to the supposed instincts of young birds and lambs, showing clearly that the former at all events are not due to inherited experience, as Darwin thought. The whole book, too, is pervaded by such a true love of nature and such a perception of its marvels and mysteries as to be unique in my experience. The modern scientific morphologists seem so wholly occupied in tracing out the mechanism of organisms that they hardly seem to appreciate the overwhelming marvel of the powers of life, which result in such infinitely varied structures and such strange habits and so-called instincts. The older I grow the more marvellous seem to me the mere variety of form and habit in plants and animals, and the unerring certitude with which from a minute germ the whole complex organism is built up, true to the type of its kind in all the infinitude of details! It is this which gives such a charm to the watching of plants growing, and of kittens so rapidly developing their senses and habitudes!...—Yours very faithfully,

ALFRED R. WALLACE.

## TO PROF. POULTON

*Parkstone, Dorset. February 1, 1893.*

My dear Poulton,—Thanks for the separate copy of your great paper on colours of larva, pupa, etc.[22] I have read your conclusions and looked over some of the experiments, and think you have now pretty well settled that question.

I am reading through the new volume of the Life of Darwin, and am struck with the curious example his own case affords of non-heredity of acquired variations. He expresses his constant dread—one of the troubles of his life—that his children would inherit his bad health. It seems pretty clear, from what F. Darwin says in the new edition, that Darwin's constant nervous stomach irritation was caused by his five years sea-sickness. It was thoroughly established before, and in the early years of, his marriage, and, on his own theory his children ought all to have inherited it. Have they? You know perhaps better than I do, whether any of the family show any symptoms of that particular form of illness—and if not it is a fine case!—Yours very faithfully,

ALFRED R. WALLACE.

Wallace was formally admitted to the Royal Society in June, 1893. The postscript of the following letter refers to his cordial reception by the Fellows.

### TO PROF. MELDOLA

*Parkstone, Dorset. June 10, 1893.*

My dear Meldola,—As we had no time to "discourse" on Thursday, I will say a few words on the individual adaptability question. We have to deal with facts, and facts certainly show that, in many groups, there is a great amount of adaptable change produced in the individual by external conditions, and that that change is not inherited. I do not see that this places Natural Selection in any subordinate position, because this individual adaptability is evidently advantageous to many species, and may itself have been produced or increased by Natural Selection. When a species is subject to great changes of conditions, either locally or at uncertain times, it may be a decided advantage to it to become individually adapted to that change while retaining the power to revert instantly to its original form when the normal conditions return. But whenever the changed conditions are permanent, or are such that individual adaptation cannot meet the requirements, then Natural Selection rapidly brings about a permanent adaptation which is inherited. In plants these two forms of adaptation are well marked and easily tested, and we shall soon have a large body of evidence upon it. In the higher animals I imagine that individual adaptation is small in amount, as indicated by the fact that even slight varieties often breed true.

In Lepidoptera we have the two forms of colour-adaptability clearly shown. Many species are, in all their stages, permanently adapted to their environment. Others have a certain power of individual adaptation, as of the pupæ to their surroundings. If this last adaptation were strictly inherited it would be positively injurious, since the progeny would thereby lose the power of individual adaptability, and thus we should have light pupæ on dark surroundings, and vice versa. Each kind of adaptation has its own sphere, and it is essential that the one should be non-inheritable, the other heritable. The whole thing seems to me quite harmonious and "as it should be."

Thiselton-Dyer tells me that H. Spencer is dreadfully disturbed on the question. He fears that acquired characters may not be inherited, in which case the foundation of his whole philosophy is undermined!—Yours very truly,

ALFRED R. WALLACE.

P.S.—I am afraid you are partly responsible for that kindly meant but too personal manifestation which disturbed the solemnity of the Royal Society meeting on Thursday!...

### TO PROF. POULTON

*Parkstone, Dorset. September 25, 1893.*

My dear Poulton,—I suppose you were not at Nottingham and did not get the letter, paper, and photographs I sent you there, but to be opened by the Secretary of Section D in case you were not there. It was about a wonderful and perfectly authenticated case of a woman who dressed the arm of a gamekeeper after amputation, and six or seven months afterwards had a child born without the forearm on the right side, exactly corresponding in *form* and *length* of stump to that of the man. Photographs of the man, and of the boy seven or eight years old, were taken *by the physician of the hospital* where the man's arm was cut off, and they show a most striking correspondence. These, with my short paper, appear to have produced an effect, for a committee of Section D has been appointed to collect evidence on this and other matters....—Yours very faithfully,

ALFRED R. WALLACE.

### TO PROF. POULTON

*Parkstone, Dorset. November 17, 1893.*

My dear Poulton,—The letter I wrote to you at Nottingham was returned to me here (after a month), so I did not think it worth while to send it to you again, though it did contain my congratulations on your appointment,[23] which I now repeat. As you have not seen the paper I sent to the British Association, I will just say that I should not have noticed the subject publicly but, after a friend had given me the photographs (sent with my paper), I came across the following statement in the new edition of Chambers' Encyclopædia, art. Deformities (by Prof. A. Hare): "In an increasing proportion of cases which are carefully investigated, it appears that maternal impressions, the result of shock or unpleasant experiences, may have a considerable influence in producing deformities in the offspring." In consequence of this I sent the case which had been furnished me, and which is certainly about as well attested and conclusive as anything can be. The facts are these:

A gamekeeper had his right forearm amputated at the North Devon Infirmary. He left before it was healed, thinking his wife could dress it, but as she was too nervous, a neighbour, a young recently married woman, a farmer's wife, still living, came and dressed it every day till it healed. About six months after she had a child born *without right hand and forearm*, the stump exactly corresponding in length to that of the gamekeeper. Dr. Richard Budd, M.D., F.R.C.P.,[24] of Barnstaple, the physician to the infirmary, when the boy was five or six years old, himself took a photograph of the boy and the gamekeeper side by side, showing the wonderful correspondence of the two arms. I have these facts *direct from Dr. Budd*, who was personally cognisant of the whole circumstances. A few years after, in November, 1876, Dr. Budd gave an account of the case and exhibited the photographs to a large meeting at the College of Physicians, and I have no doubt it is *one* of the cases referred to in the article I have quoted, though Dr. Budd thinks it has never been published. It will be at once admitted that this is not a chance coincidence, and that all theoretical difficulties must give way to such facts as this, ... Of course it by no means follows that similar causes should in all cases produce similar effects, since the idiosyncrasy of the mother is no doubt an important factor; but where the combined coincidences are so numerous as in this case—*place, time, person* and exact correspondence of *resulting deformity*—some causal relation must exist.—Believe me yours very truly,

ALFRED R. WALLACE.

## III.—Correspondence on Biology, Geographical Distribution, etc.
### [1894—1913]

### HERBERT SPENCER TO A.R. WALLACE

*Queen's Hotel, Cliftonville, Margate. August 10, 1894.*

Dear Mr. Wallace,—Though we differ on some points we agree on many, and one of the points on which we doubtless agree is the absurdity of Lord Salisbury's representation of the process of Natural Selection based upon the improbability of two varying individuals meeting. His nonsensical representation of the theory ought to be exposed, for it will mislead very many people. I see it is adopted by the *Pall Mall*. I have been myself strongly prompted to take the matter up, but it is evidently your business to do that.

Pray write a letter to the *Times* explaining that selection or survival of the fittest does not necessarily take place in the way he describes. You might set out by remarking that whereas he begins by comparing himself to a volunteer colonel reviewing a regiment of regulars, he very quickly changes his attitude and becomes a colonel of regulars reviewing volunteers and making fun of their bunglings. He deserves a-severe castigation. There are other points on which his views should be rectified, but this is the essential point.

It behoves you of all men to take up the gauntlet he has thrown down.— Very truly yours,

HERBERT SPENCER.

## HERBERT SPENCER TO A.R. WALLACE

*Queen's Hotel, Cliftonville, Margate, Aug. 19, 1894.*

Dear Mr. Wallace,—I cannot at all agree with you respecting the relative importance of the work you are doing and that which I wanted you to do. Various articles in the papers show that Lord Salisbury's argument is received with triumph, and, unless it is disposed of, it will lead to a public reaction against the doctrine of evolution at large, a far more serious evil than any error which you propose to rectify among biologists. Everybody will look to you for a reply, and if you make no reply it will be understood that Lord Salisbury's objection is valid. As to the non-publication of your letter in the *Times*, that is absurd, considering that your name and that of Darwin are constantly coupled together.—Truly yours,

HERBERT SPENCER.

## TO PROF. POULTON

*Parkstone, Dorset. September 8, 1894.*

My dear Poulton,—I was glad to see your exposure of another American Neo-Lamarckian in *Nature*.[25] It is astonishing how utterly illogical they all are! I was much pleased with your point of the adaptations supposed to be produced by the inorganic environment when they are related to the organic. It is I think new and very forcible. For nearly a month I have been wading through Bateson's book,[26] and writing a criticism of it, and of Galton, who backs him up with his idea of "organic stability." ... Neither he nor Galton appears to have any adequate conception of what Natural Selection is, or how impossible it is to escape from it. They seem to think that, given a stable variation, Natural Selection must hide its diminished head!

Bateson's preface, concluding reflections, etc., are often quite amusing.... He is so cocksure he has made a great discovery—which is the most palpable of mare's nests.—Yours very truly,

ALFRED R. WALLACE.

P.S.—I allude of course to his grand argument—"environment *continuous*—species *discontinuous*—therefore *variations* which produce species must be also *discontinuous*"! (Bateson—Q.E.D.).

## TO PROF. POULTON

*Parkstone, Dorset. February 19, 1895.*

My dear Poulton,—I have read your paper on "Theories of Evolution"[27] with great pleasure. It is very clear and very forcible, and I should think must have opened the eyes of some of your hearers. Your cases against Lamarckism were very strong, and I think quite conclusive. There is one, however, which seems to me weak—that about the claws of lobsters and the tails of lizards moving and acting when detached from the body. It may be argued, fairly, that this is only an incidental result of the extreme muscular irritability and contractibility of the organs, which might have been caused on Lamarckian as well as on the Darwinian hypothesis. The running of a fowl after its head is chopped off is an example of the same kind of thing, and this is certainly not useful. The detachment itself of claw and tail is no doubt useful and adaptive.

When discussing the objection as to failures not being found fossil, there are two additional arguments to those you adduce: (1) Every failure has been, first, a success, or it could not have come into existence (as a species); and (2) the hosts of huge and very specialised animals everywhere recently extinct are clearly failures. They were successes as long as the struggle was with animal competitors only, physical conditions being highly favourable. But, when physical conditions became adverse, as by drought, cold, etc., they failed and became extinct. The entrance of new enemies from another area might equally render them failures. As to your question about myself and Darwin, I had met him once only for a few minutes at the British Museum before I went to the East.... —Yours very faithfully,

A.R. WALLACE.

## TO MR. CLEMENT REID

*Parkstone, Dorset. November 18, 1894.*

My dear Clement Reid,— ... The great, the grand, and long-expected, the prophesied discovery has at last been made—Miocene or Old Pliocene Man in India!!! Good worked flints found *in situ* by the palæontologist to the Geological Survey of India! It is in a ferruginous conglomerate lying beneath 4,000 feet of Pliocene strata and containing hippotherium, etc. But perhaps you have seen the article in *Natural Science* describing it, by Rupert Jones, who, very properly, accepts it! Of course we want the bones, but we have got the flints, and they may follow. Hurrah for the missing link! Excuse more.— Yours very faithfully,

ALFRED R. WALLACE.


The next letter relates to the rising school of biologists who, in opposition to Darwin's views, held that species might arise by what was at the time termed "discontinuous variation."


## TO PROF. MELDOLA

*February 4, 1895.*

My dear Professor Meldola,—I hope to have copies of my "Evolution" article in a few days, and will send you a couple. The article was in print last September, but, being long, was crowded out month after month, and only now got in by being cut in two. I think I have demolished "discontinuous variation" as having any but the most subordinate part in evolution of species.

Congratulations on Presidency of the Entomological Society.

A.R. WALLACE.


## TO PROF. POULTON

*Parkstone, Dorset. March 15, 1895.*

My dear Poulton,—I have now nearly finished reading Romanes, but do not find it very convincing. There is a large amount of special pleading. On two points only I feel myself hit. My doubt that Darwin really meant that *all* the individuals of a species could be similarly modified without selection is evidently wrong, as he adduces other quotations which I had overlooked. The other point is, that my suggested explanation of sexual ornaments gives away my case as to the utility of all specific characters. It certainly does as it stands, but I now believe, and should have added, that all these ornaments, where they differ from species to species, are also recognition characters, and as such were rendered stable by Natural Selection from their first appearance.

I rather doubt the view you state, and which Gulick and Romanes make much of, that a portion of a species, separated from the main body, will have a different average of characters, unless they are a local race which has already been somewhat selected. The large amount of variation, and the regularity of the curve of variation, whenever about 50 or 100 individuals are measured in the same locality, shows that the bulk of a species are similar in amount of variation everywhere. But when a portion of a species begins to be modified in adaptation to new conditions, distinction of some kind is essential, and therefore any slight difference would be increased by selection. I see no reason to believe that species (usually) have been isolated first and modified afterwards, but rather that new species usually arise from species which have a wide range, and in different areas need somewhat different characters and habits. Then *distinctness* arises both by adaptation and by development of recognition marks to minimise intercrossing.

I wonder Darwin did not see that if the unknown "constant causes" he supposes can modify all the individuals of a species, either indifferently, usefully, or hurtfully, and that these characters so produced are, as Romanes says, very, very numerous in all species, and are sometimes the only specific characters, then the Neo-Lamarckians are quite right in putting Natural Selection as a very secondary and subordinate influence, since all it has to do is to weed out the hurtful variations.

Of course, if a species with warning colours were, in part, completely isolated, and its colours or markings were accidentally different from the parent form, whatever set of markings and colours it had would be, I consider, rendered stable for recognition, and also for protection, since if it varied too much the young birds and other enemies would take a heavier toll in learning it was uneatable. It might then be said that the character by which this species differs from the parent species is a useless character. But surely this is not what is usually meant by a "useless character." This is highly useful in itself, though the difference from the other species is not useful. If they were in contact it would be useful, as a distinction preventing intercrossing, and so long as they are not brought together we cannot really tell if it is a species at all, since it might breed freely with the parent form and thus return back to one type. The "useless characters" I have always had in mind when arguing this question are those which are or are supposed to be absolutely useless, not merely relatively as regards the difference from an allied species. I think this is an important distinction.—Yours very truly,

ALFRED R. WALLACE.

## HERBERT SPENCER TO A.R. WALLACE

*64 Avenue Road, Regent's Park, London, N.W. September 28, 1895*

Dear Mr. Wallace,—As I cannot get you to deal with Lord Salisbury I have decided to do it myself, having been finally exasperated into doing it by this honour paid to his address in France—the presentation of a translation to the French Academy. The impression produced upon some millions of people in England cannot be allowed to be thus further confirmed without protest.

One of the points which I propose to take up is the absurd conception Lord Salisbury sets forth of the process of Natural Selection. When you wrote you said you had dealt with it yourself in your volume on Darwinism. I have no doubt that it is also in some measure dealt with by Darwin himself, by implication or incidentally. You of course know Darwin by heart, and perhaps you would be kind enough to save me the trouble of searching by indicating the relevant passages both in his books and in your own. My reading power is very small, and it tries me to find the parts I want by much reading.—Truly yours,

HERBERT SPENCER.


To the following letter from Mr. Gladstone, Wallace attached this pencil note: "In 1881 I put forth the first idea of mouth-gesture as a factor in the origin of language, in a review of E.B. Tylor's 'Anthropology,' and in 1895 I extended it into an article in the *Fortnightly Review*, and reprinted it with a few further corrections in my 'Studies,' under the title 'The Expressiveness of Speech or Mouth-Gesture as a Factor in the Origin of Language.' In it I have developed a completely new principle in the theory of the origin of language by showing that every motion of the jaws, lips and tongue, together with inward or outward breathing, and especially the mute or liquid consonants ending words which serve to indicate abrupt or continuous motion, have corresponding meanings in so many cases as to show a fundamental connection. I thus enormously extended the principle of onomatopoeia in the origin of vocal language. As I have been unable to find any reference to this important factor in the origin of language, and as no competent writer has pointed out any fallacy in it, I think I am justified in supposing it to be new and important. Mr. Gladstone informed me that there were many thousands of illustrations of my ideas in Homer."—A.R.W.

## W.E. GLADSTONE TO A.R. WALLACE

*Hawarden Castle, Chester. October 18, 1895.*

Dear Sir,—Your kindness in sending me your most interesting article draws on you the inconvenience of an acknowledgment.

My pursuits in connection with Homer, especially, have made me a confident advocate of the doctrine that there is, within limits, a connection in language between sound and sense.

I would consent to take the issue simply on English words beginning with *st*. You go upon a kindred class in *sn*. I do not remember a perfectly *innocent* word, a word habitually used *in bonam partem*, and beginning with *sn*, except the word "snow," and "snow," as I gather from *Schnee*, is one of the worn-down words.

May I beg to illustrate you once more on the ending in *p*. I take our old schoolboy combinations: hop, skip and jump. Each motion an ending motion; and to each word closed with *p* compare the words *run, rennen, courir, currere*.

But I have now a new title to speak. It is deafness; and I know from deafness that I run a worse chance with a man whose mouth is covered with beard and moustache.

A young relation of mine, slightly deaf, was sorely put to it in an University examination because one of his examiners was *secretal* in this way.

I will not trouble you further except to express, with misgiving, a doubt on a single point, the final *f*.

In driving with Lord Granville, who was deaf but not very deaf, I had occasion to mention to him the Duke of *Fife*, I used every effort, but in no way could I contrive to make him hear the word.

I break my word to add one other particular. Out of 27,000 odd lines in Homer, every one of them expressed, in a sense, heavy weight or force; the blows of heavy-armed men on the breastplates of foes ... [illegible] and the like.—With many thanks, I remain yours very faithfully,

W.E. GLADSTONE.

P.S.—I should say that the efficacy of lip-expression, undeniably, is most subtle, and defies definite description.

## TO DR. ARCHDALL REID

*Parkstone, Dorset. April 19, 1896.*

Dear Sir,—I am sorry I had not space to refer more fully to your interesting work.[28] The most important point on which I think your views require emendation is on *instinct*. I see you quote Spalding's experiments, but these have been quite superseded and shown to be seriously incorrect by Prof. Lloyd Morgan. A paper by him in the *Fortnightly Review* of August, 1893, gives an account of his experiments, and he read a paper on the same subject at the British Association last year. He is now preparing a volume on the subject which will contain the most valuable series of observations yet made on this question. Another point of some importance where I cannot agree with you is your treating dipsomania as a disease, only to be eliminated by drunkenness and its effects. It appears to me to be only a vicious habit or indulgence which would cease to exist in a state of society in which the habit were almost universally reprobated, and the means for its indulgence almost absent. But this is a matter of comparatively small importance.—Believe me yours very truly,

ALFRED R. WALLACE.

## TO DR. ARCHDALL REID

*Parkstone. April 28, 1896.*

Dear Sir,—"We can but reason from the facts we know." We know a good deal of the senses of the higher animals, very little of those of insects. If we find—as I think we do—that all cases of supposed "instinctive knowledge" in the former turn out to be merely intuitive reactions to various kinds of stimulus, combined with very rapidly acquired experience, we shall be justified in thinking that the actions of the latter will some day be similarly explained. When Lloyd Morgan's book is published we shall have much information on this question. (*See* "Natural Selection and Tropical Nature," pp. 91-7.)—Yours truly, ALFRED R. WALLACE.

## TO PROF. MELDOLA

*Parkstone, Dorset. October 12, 1896.*

My dear Meldola,—I got Weismann's "Germinal Selection" two or three months back and read it very carefully, and on the whole I admire it very much, and think it does complete the work of ordinary variation and selection. Of course it is a pure hypothesis, and can never perhaps be directly proved, but it seems to me a reasonable one, and it enables us to understand two groups of facts which I have never been able to work out satisfactorily by the old method. These two facts are: (1) the total, or almost total, disappearance of many useless organs, and (2) the continuous development of secondary sexual characters beyond any conceivable utility, and, apparently, till checked by inutility. It explains both these. Disuse alone, as I and many others have always argued, cannot do the first, but can only cause *regression to the mean*, with perhaps some further regression from economy of material.

As to the second, I have always felt the difficulty of accounting for the enormous development of the peacock's train, the bird of paradise plumes, the long wattle of the bell bird, the enormous tail-feathers of the Guatemalan trogon, of some humming-birds, etc. etc. etc. The beginnings of all these I can explain as recognition marks, and this explains also their distinctive character in allied species, but it does not explain their growing on and on far beyond what is needful for recognition, and apparently till limited by absolute hurtfulness. It is a relief to me to have "germinal selection" to explain this.

I do not, however, think it at all necessary to explain adaptations, however complex. Variation is so general and so large, in dominant species, and selection is so tremendously powerful, that I believe all needful adaptation may be produced without it. But, if it exists, it would undoubtedly hasten the process of such adaptation and would therefore enable new places in the economy of nature to be more rapidly filled up.

I was thinking of writing a popular exposition of the new theory for *Nature*, but have not yet found time or inclination for it. I began reading "Germinal Selection" with a prejudice against it. That prejudice continued through the first half, but when I came to the idea itself, and after some trouble grasped the meaning and bearing of it, I saw the work it would do and was a convert at once. It really has no relation to Lamarckism, and leaves the non-heredity of acquired characters exactly where it was.—Yours very truly,

ALFRED R. WALLACE.

The next letter relates to the great controversy then being carried on with respect to Weismann's doctrine of the non-inheritance of "acquired" characters, which doctrine implied complete rejection of the last trace of Lamarckism from Darwinian evolution. Wallace ultimately accepted the Weismannian teaching. Darwin had no opportunity during his lifetime of considering this question, which was raised later in an acute form by Weismann.

## TO PROF. MELDOLA

*Parkstane, Dorset. January 6, 1897.*

My dear Meldola,—The passage to which you refer in the "Origin" (top of p. 6) shows Darwin's firm belief in the "heredity of acquired variations," and also in the importance of definite variations, that is, "sports," though elsewhere he almost gives these up in favour of indefinite variations; and this last is now the view of all Darwinians, and even of many Lamarckians. I therefore always now assume this as admitted. Weismann's view as to "possible variations" and "impossible variations" on p. 1 of "Germinal Selection" is misleading, because it can only refer to "sports" or to "cumulative results," not to "individual variations" such as are the material Natural Selection acts on. Variation, as I understand it, can only be a slight modification in the offspring of that which exists in the parent. The question whether pigs could possibly develop wings is absurd, and altogether beside the question, which is, solely, so far as direct evidence goes, as to the means by which the change from one species to another closely allied species has been brought about. Those who want to begin by discussing the causes of change from a dog to a seal, or from a cow to a whale, are not worth arguing with, as they evidently do not comprehend the A, B, C of the theory.

Darwin's ineradicable acceptance of the theory of heredity of the effects of climate, use and disuse, food, etc., on the individual led to much obscurity and fallacy in his arguments, here and there.—Yours very sincerely,

ALFRED R. WALLACE.


## TO PROF. POULTON

*Parkstone, Dorset. February 14, 1897.*

My dear Poulton,—Thanks for copy of your British Association Address,[29] which I did not read in *Nature*, being very busy just then. I have now read it with much pleasure, and think it a very useful and excellent discussion that was much needed. There is, however, one important error, I think, which vitiates a vital part of the argument, and which renders it possible so to reduce the time indicated by geology as to render the accordance of Geology and Physics more easy to effect. The error I allude to was made by Sir A. Geikie in his Presidential Address[30] which you quote. Immediately it appeared I wrote to him pointing it out, but he merely acknowledged my letter, saying he would consider it. To me it seems a most palpable and extraordinary blunder. The error consists in taking the rate of deposition as the same as the rate of denudation, whereas it is about twenty times as great, perhaps much more—because the area of deposition is at least twenty times less than that of denudation. In order to equal the area of denudation, it would require that *every* bed of *every* formation should have once extended over the *whole area* of all the land of the globe! The deposition in narrow belts along coasts of all the matter brought down by rivers, as proved by the *Challenger*, leads to the same result. In my "Island Life," 2nd Edit., pp. 221-225, I have discussed this whole matter, and on reading it again I can find no fallacy in it. I have, however, I believe, overestimated the time required for deposition, which I believe would be more nearly one-fortieth than one-twentieth that of mean denudation; because there is, I believe, also a great overestimate of the maximum of deposition, because it is partly made up of beds which may have been deposited simultaneously. Also the maximum thickness is probably double the mean thickness.

The mean rate of denudation, both for European rivers and for all the rivers that have been measured, is a foot in three million years, which is the figure that should be taken in calculations.—Believe me yours very truly,

ALFRED R. WALLACE.

## TO PROF. MELDOLA

*Parkstone, Dorset. April 27, 1897.*

My dear Meldola,— ... I thought Romanes' article in reply to Spencer was very well written and wonderfully clear for him, and I agree with most of it, except his high estimate of Spencer's co-adaptation argument. It is quite true that Spencer's biology rests entirely on Lamarckism, so far as heredity of acquired characters goes. I have been reading Weismann's last book, "The Germ Plasm." It is a wonderful attempt to solve the most complex of all problems, and is almost unreadable without some practical acquaintance with germs and their development.—Believe me yours very faithfully,

ALFRED R. WALLACE.

## TO PROF. POULTON

*Parkstone, Dorset. June 13, 1897.*

My dear Poulton,— ... The rate of deposition might be modified in an archipelago, but would not necessarily be less than now, on the *average*. On the ocean side it might be slow, but wherever there were comparatively narrow straits between the islands it might be even faster than now, because the area of deposition would be strictly limited. In the seas between Java and

Borneo and between Borneo and Celebes the deposition *may be* above the average. Again, during the development of continents there were evidently extensive mountain ridges and masses with landlocked seas, or inland lakes, and in all these deposition would be rapid. Anyhow, the fact remains that there is no necessary equality between rates of denudation and deposition (in thickness) as Geikie has *assumed*.

I was delighted with your account of Prichard's wonderful anticipation of Galton and Weismann! It is so perfect and complete.... It is most remarkable that such a complete statement of the theory and such a thorough appreciation of its effects and bearing should have been so long overlooked. I read Prichard when I was very young, and have never seen the book since. His facts and arguments are really useful ones, and I should think Weismann must be delighted to have such a supporter come from the grave. His view as to the supposed transmission of disease is quite that of Archdall Reid's recent book. He was equally clear as to Selection, and had he been a *zoologist* and *traveller* he might have anticipated the work of both Darwin and Weismann!

To bring out such a book as his "Researches" when only twenty-seven, and a practising physician, shows what a remarkable man he was.—Believe me yours very truly,

ALFRED R. WALLACE.

## TO PROF. MELDOLA

*Parkstone, Dorset. July 8, 1897.*

My dear Meldola,— ... I am now reading a wonderfully interesting book— O. Fisher's "Physics of the Earth's Crust." It is really a grand book, and, though full of unintelligible mathematics, is so clearly explained and so full of good reasoning on all the aspects of this most difficult question that it is a pleasure to read it. It was especially a pleasure to me because I had just been writing an article on the Permanence of the Oceanic Basins, at the request of the Editor of *Natural Science*, who told me I was not orthodox on the point. But I find that Fisher supports the same view with very great force, and it strikes me that if weight of argument and number of capable supporters create orthodoxy in science, it is the other side who are not orthodox. I have some fresh arguments, and I was delighted to be able to quote Fisher. It seems almost demonstrated now that Sir W. Thomson was wrong, and that the earth *has* a molten interior and a very thin crust, and in no other way can the phenomena of geology be explained....—Yours very truly,

ALFRED R. WALLACE.

## TO SIR OLIVER LODGE

*Parkstone, Dorset. March 8, 1898.*

My dear Sir,—My own opinion has long been—and I have many times given reasons for it—that there is always an ample amount of variation in all directions to allow any useful modification to be produced, very rapidly, as compared with the rate of those secular changes (climate and geography) which necessitate adaptation; hence no guidance of variation in certain lines is necessary. For proof of this I would ask you to look at the diagrams in Chapter III. of my "Darwinism," reading the explanation in the text. The proof of such constant indefinite variability has been much increased of late years, and if you consider that instead of tens or hundreds of individuals, Nature has as many thousands or millions to be selected from, every year or two, it will be clear that the materials for adaptation are ample.

Again, I believe that the time, even as limited by Lord Kelvin's calculations, is ample, for reasons given in Chapter X., "On the Earth's Age," in my "Island Life," and summed up on p. 236. I therefore consider the difficulty set forth on p. 2 of the leaflet you send is not a real one. To my mind, the development of plants and animals from low forms of each is fully explained by the variability proved to exist, with the actual rapid multiplication and Natural Selection. For this no other intellectual agency is required. The problem is to account for the infinitely complex constitution of the material world and its forces which rendered living organisms possible; then, the introduction of consciousness or sensation, which alone rendered the animal world possible; lastly, the presence in man of capacities and moral ideas and aspirations which could not conceivably be produced by variation and Natural Selection. This is stated at p. 473-8 of my "Darwinism," and is also referred to in the article I enclose (at p. 443) and which you need not return.

The subject is so large and complex that it is not to be wondered so many people still maintain the insufficiency of Natural Selection, without having really mastered the facts. I could not, therefore, answer your question without going into some detail and giving references.... —Believe me yours very truly,

ALFRED R. WALLACE.

## TO MR. H.N. RIDLEY

*Parkstone, Dorset. October 3, 1898.*

My dear Mr. Ridley,— ... We are much interested now about De Rougemont, and I dare say you have seen his story in the *Wide World Magazine*, while in the *Daily Chronicle* there have been letters, interviews and discussions without end. A few people, who think they know everything, treat him as an impostor; but unfortunately they themselves contradict each other, and so far are proved to be wrong more often than De Rougemont. I firmly believe that his story is substantially true—making allowance for his being a foreigner who learnt one system of measures, then lived thirty years among savages, and afterwards had to reproduce all his knowledge in English and Australian idioms. As an intelligent writer in the *Saturday Review* says, putting aside the sensational illustrations there is absolutely nothing in his story but what is quite *possible* and even *probable*. He must have reached Singapore the year after I returned home, and I dare say there are people there who remember Jensen, the owner of the schooner *Veilland*, with whom he sailed on his disastrous pearl-fishing expedition. Jensen is said now to be in British New Guinea, and has often spoken of his lost cargo of pearls. —— and ——, of the Royal Geographical Society, state that they are convinced of the substantial truth of the main outlines of his story, and after three interviews and innumerable questions are satisfied of his *bona fides*—and so am I.—With best wishes, believe me to be yours very truly,

ALFRED R. WALLACE.

## MR. SAMUEL WADDINGTON TO A.R. WALLACE

*7 Whitehall Gardens, London, S.W. February 19, 1901.*

Dear Sir,—I trust you will forgive a stranger troubling you with a letter, but a friend has asked me whether, as a matter of fact, Darwin held that *all* living creatures descended from one and the same ancestor, and that the pedigree of a humming-bird and that of a hippopotamus would meet if traced far enough back. Can you tell me whether Darwin did teach this?

I should have thought that as life was developed once, it probably could and would be developed many times in different places, as month after month, and year after year went by; and that, from the very first, it probably took many different forms and characters, in the same way as crystals take different forms and shapes, even when composed of the same substance. From these many developments of "life" would descend as many separate lines of evolution, one ending in the humming-bird, another in the hippopotamus, a third in the kangaroo, etc., and their pedigrees (however far back they might be traced) would not join until they reached some primitive form of protoplasm,—Yours faithfully,

SAMUEL WADDINGTON.

## TO MR. SAMUEL WADDINGTON

*Parkstone, Dorset. February 23, 1901.*

Dear Sir,—Darwin believed that all living things originated from "a few forms or from one"—as stated in the last sentence of his "Origin of Species." But privately I am sure he believed in the *one* origin. Of course there is a possibility that there were several distinct origins from inorganic matter, but that is very improbable, because in that case we should expect to find some difference in the earliest forms of the germs of life. But there is no such difference, the primitive germ-cells of man, fish or oyster being almost indistinguishable, formed of identical matter and going through identical primitive changes.

As to the humming-bird and hippopotamus, there is no doubt whatever of a common origin—if evolution is accepted at all; since both are vertebrates—a very high type of organism whose ancestral forms can be traced back to a simple type much earlier than the common origin of mammals, birds and reptiles.—Yours very truly,

ALFRED R. WALLACE.

## TO SIR FRANCIS DARWIN

*Parkstone, Dorset. July 3, 1901.*

Dear Mr. Darwin,—Thanks for the letter returned. I *do* hold the opinion expressed in the last sentence of the article you refer to, and have reprinted it in my volume of Studies, etc. But the stress must be laid on the word *proof.* I intended it to enforce the somewhat similar opinion of your father, in the "Origin" (p. 424, 6th Edit.), where he says, "Analogy may be a deceitful guide." But I really do not go so far as he did. For he maintained that there was not any proof that the several great classes or kingdoms were descended from common ancestors.

I maintain, on the contrary, that all without exception are now proved to have originated by "descent with modification," but that there is no proof, and no necessity, that the very same causes which have been sufficient to produce all the species of a genus or Order were those which initiated and developed the greater differences. At the same time I do *not* say they were not sufficient. I merely urge that there is a difference between proof and probability.—Yours very truly,

ALFRED R. WALLACE.

### TO PROF. POULTON

*Broadstone, Wimborne. August 5, 1904.*

My dear Poulton,— ... What a miserable abortion of a theory is "Mutation," which the Americans now seem to be taking up in place of Lamarckism, "superseded." Anything rather than Darwinism! I am glad Dr. F.A. Dixey shows it up so well in this week's *Nature*,[31] but too mildly!—Yours very truly,

ALFRED R. WALLACE.

### TO PROF. POULTON

*Broadstone, Wimborne. April 3, 1905.*

My dear Poulton,—Many thanks for copy of your Address,[32] which I have read with great pleasure and will forward to Birch next mail. You have, I think, produced a splendid and unanswerable set of facts proving the non-heredity of acquired characters. I was particularly pleased with the portion on "instincts," in which the argument is especially clear and strong. I am afraid, however, the whole subject is above and beyond the average "entomologist" or insect collector, but it will be of great value to all students of evolution. It is curious how few even of the more acute minds take the trouble to reason out carefully the teaching of certain facts—as in the case of Romanes and the "variable protection," and as I showed also in the case of Mivart (and also Romanes and Gulick) declaring that isolation alone, without Natural Selection, could produce perfect and well-defined species (see *Nature*, Jan. 12, 1899).... —Yours faithfully,

A.R. WALLACE.

## TO SIR FRANCIS DARWIN

*Broadstone, Wimborne. October 29, 1905.*

Dear Mr. Darwin,—I return you the two articles on "Mutation" with many thanks. As they are both supporters of de Vries, I suppose they put his case as strongly as possible. Professor Hubrecht's paper is by far the clearest and the best written, and he says distinctly that de Vries claims that all new species have been produced by mutations, and none by "fluctuating variations." Professor Hubrecht supports this and says that de Vries has proved it! And all this founded upon a few "sports" from one species of plant, itself of doubtful origin (variety or hybrid), and offering phenomena in no way different from scores of other cultivated plants. Never, I should think, has such a vast hypothetical structure been erected on so flimsy a basis!

The boldness of his statements is amazing, as when he declares (as if it were a fact of observation) that fluctuating variability, though he admits it as the origin of all domestic animals and plants, yet "never leads to the formation of species"! (Hubrecht, p. 216.) There is one point where he so grossly misinterprets your father that I think you or some other botanist should point it out. De Vries is said to quote from "Life and Letters," II., p. 83, where Darwin refers to "chance variations"—explained three lines on as "the slight differences selected by which a race or species is at length formed." Yet de Vries and Hubrecht claim that by "chance variations" Darwin meant "sports" or "mutations," and therefore agrees with de Vries, while both omit to refer to the many passages in which, later, he gave less and less weight to what he termed "single large variations"—the same as de Vries' "mutations"!—Yours very truly,

ALFRED R. WALLACE.

## TO SIR JOSEPH HOOKER

*Broadstone, Wimborne. November 10, 1905.*

My dear Sir Joseph,—I am writing to apologise for a great oversight. When I sent my publishers a list of persons who had contributed to "My Life" in various ways, your name, which should have been *first*, was strangely omitted, and the omission was only recalled to me yesterday by reading your letters to Bates in Clodd's edition of his Amazon book, which I have just purchased. I now send you a copy by parcel-post, in the hope that you will excuse the omission to send it sooner.

Now for a more interesting subject, I was extremely pleased and even greatly surprised, in reading your letters to Bates, to find that at that early period (1862) you were already strongly convinced of three facts which are absolutely essential to a comprehension of the method of organic evolution, but which many writers, even now, almost wholly ignore. They are (1) the universality and large amount of normal variability, (2) the extreme rigour of Natural Selection, and (3) that there is no adequate evidence for, and very much against, the inheritance of acquired characters.

It was only some years later, when I began to write on the subject and had to think out the exact mode of action of Natural Selection, that I myself arrived at (1) and (2), and have ever since dwelt upon them—in season and out of season, as many will think—as being absolutely essential to a comprehension of organic evolution. The third I did not realise till I read Weismann, I have never seen the sufficiency of normal variability for the modification of species more strongly or better put than in your letters to Bates. Darwin himself never realised it, and consequently played into the hands of the "discontinuous variation" and "mutation" men, by so continually saying "*if* they vary"—"without variation Natural Selection can do nothing," etc.

Your argument that variations are not caused by change of environment is equally forcible and convincing. Has anybody answered de Vries yet?

F. Darwin lent me Prof. Hubrecht's review from the *Popular Science Monthly*, in which he claims that de Vries has proved that new species have always been produced from "mutations," never through normal variability, and that Darwin latterly agreed with him! This is to me amazing! The Americans too accept de Vries as a second Darwin!—Yours very sincerely,

ALFRED E. WALLACE.

## SIR J. HOOKER TO A.R. WALLACE

*The Camp, Sunningdale. November 12, 1905.*

My dear Wallace,—My return from a short holiday at Sidmouth last Thursday was greeted by your kind and welcome letter and copy of your "Life." The latter was, I assure you, never expected, knowing as I do the demand for free copies that such a work inflicts on the writer. In fact I had put it down as one of the annual Christmas gifts of books that I receive from my own family. Coming, as it thus did, quite unexpectedly, it is doubly welcome, and I do heartily thank you for this proof of your greatly valued friendship. It will prove to be one of four works of greatest interest to me of any published since Darwin's "Origin," the others being Waddell's "Lhasa," Scott's "Antarctic Voyage," and Mill's "Siege of the South Pole."

I have not seen Clodd's edition of Bates's "Amazon," which I have put down as to be got, and I had no idea that I should have appeared in it. Your citation of my letters and their contents are like dreams to me; but to tell you the truth, I am getting dull of memory as well as of hearing, and what is worse, in reading: what goes in at one eye goes out at the other. So I am getting to realise Darwin's consolation of old age, that it absolves me from being expected to know, remember, or reason upon new facts and discoveries. And this must apply to your query as to anyone having as yet answered de Vries. I cannot remember having seen any answer; only criticisms of a discontinuous sort. I cannot for a moment entertain the idea that Darwin ever assented to the proposition that new species have always been produced from mutation and never through normal variability. Possibly there is some quibble on the definition of mutation or of variation. The Americans are prone to believe any new things, witness their swallowing the thornless cactus produced by that man in California—I forget his name—which Kew exposed by asking for specimens to exhibit in the Cactus House....—I am, my dear Wallace, sincerely yours,

JOS. D. HOOKER.

## TO MR. E. SMEDLEY

*Broadstone, Wimborne. January 31, 1906.*

Dear Mr. Smedley,—I have read Oliver Lodge's book in answer to Haeckel, but I do not think it very well done or at all clearly written or well argued. A book[33] has been sent me, however, which is a masterpiece of clearness and sound reasoning on such difficult questions, and is a far more crushing reply to Haeckel than O. Lodge's. I therefore send you a copy, and feel sure you will enjoy it. It is a stiff piece of reasoning, and wants close attention and careful thought, but I think you will be able to appreciate it. In my opinion it comes as near to an intelligible solution of these great problems of the Universe as we are likely to get while on earth. It is a book to read and think over, and read again. It is a masterpiece....—Yours very truly,

ALFRED R. WALLACE.

## TO PROF. POULTON

*Broadstone, Wimborne. July 27, 1907.*

My dear Poulton,—Thanks for your very interesting letter. I am glad to hear you have a new book on "Evolution"[34] nearly ready and that in it you will do something to expose the fallacies of the Mutationists and Mendelians, who pose before the world as having got *all* wisdom, before which we poor Darwinians must hide our diminished heads!

Wishing to know the best that could be said for these latter-day anti-Darwinians, I have just been reading Lock's book on "Variation, Heredity, and Evolution." In the early part of his book he gives a tolerably fair account of Natural Selection, etc. But he gradually turns to Mendelism as the "one thing needful"—stating that there can be "no sort of doubt" that Mendel's paper is the "most important" contribution of its size ever made to biological science!

"Mutation," as a theory, is absolutely nothing new—only the assertion that new species originate *always* in sports, for which the evidence adduced is the most meagre and inconclusive of any ever set forth with such pretentious claims! I hope you will thoroughly expose this absurd claim.

Mendelism is something new, and within its very limited range, important, as leading to conceptions as to the causes and laws of heredity, but only misleading when adduced as the true origin of species in nature, as to which it seems to me to have no part.—Yours very truly,

ALFRED R. WALLACE.


## TO PROF. POULTON

*Broadstone, Wimborne. November 26, 1907.*

My dear Poulton,—Many thanks for letting me see the proofs.[35] ... The whole reads very clearly, and I am delighted with the way you expose the Mendelian and Mutational absurd claims. That ought to really open the eyes of the newspaper men to the fact that Natural Selection and Darwinism are not only holding their ground but are becoming more firmly established than ever by every fresh research into the ways and workings of living nature. I shall look forward to great pleasure in reading the whole book. I was greatly pleased with Archdall Reid's view of Mendelism in *Nature*.[36] He is a very clear and original thinker.

I see in Essay X. you use in the title the term "defensive coloration." Why this instead of the usual "protective"? Surely the whole function of such colours and markings is to protect from attack—not to defend when attacked. The latter is the function of stings, spines and hard coats. I only mention this because using different terms may lead to some misconception.

Your illustration of mutation by throwing colours on a screen, and the argument founded on it, I liked much. That reminds me that H. Spencer's argument for inheritance of acquired variations—that co-ordination of many parts at once, required for adaptations, would be impossible by chance variations of those parts—applies with a hundredfold force to mutations, which are admittedly so much less frequent both in their numbers and the repetitions of them.—Yours very truly,

ALFRED R. WALLACE.

## TO PROF. POULTON

*Broadstone, Wimborne. December 18, 1907.*

My dear Poulton,—The importance of Mendelism to Evolution seems to me to be something of the same kind, but very much less in degree and importance, as Galton's fine discovery of the law of the average share each parent has in the characters of the child—one quarter, the four grandparents each one-sixteenth, and so on. That illuminates the whole problem of heredity, combined with individual diversity, in a way nothing else does. I almost wish you could introduce that!—Yours very truly,

ALFRED R. WALLACE.

## TO DR. ARCHDALL REID

*Broadstone, Wimborne. January 19, 1908.*

Dear Sir,— ... I was much pleased the other day to read, in a review of Mr. T. Rice Holmes's fine work on "Ancient Britain and the Invasions of Julius Cæsar," that the author has arrived by purely historical study at the conclusion that we have not risen morally above our primitive ancestors. It is a curious and important coincidence.

I myself got the germ of the idea many years ago, from a very acute thinker, Mr. Albert Mott, who gave some very original and thoughtful addresses as President of the Liverpool Philosophical Society, one of which dealt with the question of savages being often, perhaps always, the descendants of more civilised races, and therefore affording no proof of progression. At that time (about 1860-70) I could not accept the view, but I have now come to think he was right.—Yours very truly,

ALFRED R. WALLACE.

## TO PROF. POULTON

*Old Orchard, Broadstone, Wimborne. November 2, 1908.*

My dear Poulton,— ... You may perhaps have heard that I have been invited by the Royal Institution (through Sir W. Crookes) to give them a lecture on the jubilee of the "Origin of Species" in January, After some consideration I accepted, because I *think* I can give a broad and general view of Darwinism, that will finally squash up the Mutationists and Mendelians, and be both generally intelligible and interesting. So far as I know this has never yet been done, and the Royal Institution audience is just the intelligent and non-specialist one I shall be glad to give it to if I can.

I have been very poorly the last three weeks, but am now recovering my health and strength slowly. It will take me all my time the next two months to get this ready, and now I must write a letter in reply to the absurd and gross misrepresentation of Prof. Hubrecht, as to imaginary differences between Darwin and myself, in the last *Contemporary*!—Yours very truly,

ALFRED R. WALLACE.

The next letter relates to Wallace's Friday evening Discourse at the Royal Institution. His friends were afraid whether his voice could be sustained throughout the hour—fears which were abundantly dispelled by the actual performance. This was his last public lecture.

## TO PROF. MELDOLA

*Old Orchard, Broadstone, Wimborne. December 20, 1908.*

My dear Meldola,—Thanks for your kind offer to read for me if necessary. But when Sir Wm. Crookes first wrote to me about it, he offered to read all, or any parts of the lecture, if my voice did not hold out. I am very much afraid I cannot stand the strain of speaking beyond my natural tone for an hour, or even for half that time—but I may be able to do the opening and conclusion....

I am glad that you see, as I do, the utter futility of the claims of the Mutationists. I may just mention them in the lecture, but I hope I have put the subject in such a way that even "the meanest capacity" will suffice to see the absurdity of their claims.—Yours very truly,

ALFRED R. WALLACE.

## TO PROF. POULTON

*Old Orchard, Broadstone, Wimborne. January 26, 1909.*

My dear Poulton,—I had a delightful two hours at the Museum on Saturday morning, as Mr. Rothschild brought from Tring several of his glass-bottomed drawers with his finest new New Guinea butterflies. They *were* a treat! I never saw anything more lovely and interesting!...

As to your very kind and pressing invitation,[1] I am sorry to be obliged to decline it. I cannot remain more than one day or night away from home, without considerable discomfort, and all the attractions of your celebration are, to me, repulsions....

My lecture, even as it will be published in the *Fortnightly*, will be far too short for exposition of all the points I wish to discuss, and I hope to occupy myself during this year in saying all I want to say in a book (of a wider scope) which is already arranged for. One of the great points, which I just touched on in the lecture, is to show that all that is usually considered the waste of Nature—the enormous number produced in proportion to the few that survive—was absolutely essential in order to secure the variety and continuity of life through all the ages, and especially of that one line of descent which culminated in man. That, I think, is a subject no one has yet dealt with.—Yours very faithfully,

ALFRED R. WALLACE.

## TO PROF. POULTON

*Old Orchard, Broadstone, Wimborne. March 1, 1909.*

Dear Poulton,— ... I am glad that Lankester has replied to the almost disgraceful Centenary article in the *Times*. But it is an illustration of the widespread mischief the Mutationists, etc., are doing. I have no doubt, however, it will all come right in the end, though the end may be far off, and in the meantime we must simply go on, and show, at every opportunity, that Darwinism actually does explain the whole fields of phenomena that they do not even attempt to deal with, or even approach....—Yours very truly,

ALFRED R. WALLACE.

## TO MRS. FISHER

*Old Orchard, Broadstone, Wimborne. March 6, 1909.*

Dear Mrs. Fisher,— ... Another point I am becoming more and more impressed with is, a teleology of fundamental laws and forces rendering development of the infinity of life-forms possible (and certain) in place of the old teleology applied to the production of each species. Such are the case of feathers reproduced annually, which I gave at end of lecture, and the still more marvellous fact of the caterpillar, often in two or three weeks of chrysalis life, having its whole internal, muscular, nervous, locomotive and alimentary organs decomposed and recomposed into a totally different being—an absolute miracle if ever there is one, quite as wonderful as would be the production of a complex marine organism out of a mass of protoplasm. Yet, because there has been continuity, the difficulty is slurred over or thought to be explained!—Yours very truly,

ALFRED R. WALLACE.

## TO SIR W.T. THISELTON-DYER

*Old Orchard, Broadstone, Wimborne. June 22, 1909.*

Dear Sir William,—On Saturday, to my great pleasure, I received a copy of the Darwin Commemoration volume. I at once began reading your most excellent paper on the Geographical Distribution of Plants. It is intensely interesting to me, both because it so clearly brings out Darwin's views and so judiciously expounds his arguments—even when you intimate a difference of opinion—but especially because you bring out so clearly and strongly his views on the general permanence of continents and oceans, which to-day, as much as ever, wants insisting upon. I may just mention here that none of the people who still insist on former continents where now are deep oceans have ever dealt with the almost physical impossibility of such a change having occurred without breaking the continuity of terrestrial life, owing to the mean depth of the ocean being at least six times the mean height of the land, and its area nearly three times, so that the whole mass of the land of the existing continents would be required to build up even *one small* continent in the depths of the Atlantic or Pacific! I have demonstrated this, with a diagram, in my "Darwinism" (Chap, XII.), and it has never been either refuted or noticed, but passed by as if it did not exist! Your whole discussion of Dispersal and Distribution is also admirable, and I was much interested with your quotations from Guppy, whose book I have not seen, but must read.

Most valuable to me also are your numerous references to Darwin's letters, so that the article serves as a compendious index to the five volumes, as regards this subject.

Especially admirable is the way in which you have always kept Darwin before us as the centre of the whole discussion, while at the same time fairly stating the sometimes adverse views of those who differ from him on certain points....—Yours very truly,

ALFRED R. WALLACE.

## SIR W.T. THISELTON-DYER TO A.R. WALLACE

*The Ferns, Witcombe, Gloucester. June 25, 1909.*

Dear Dr. Wallace,—It is difficult for me to tell you how gratified I am by your extraordinarily kind letter.... The truth is that success was easy. It has been my immense good fortune to know most of those who played in the drama. The story simply wanted a straightforward amanuensis to tell itself. But it is a real pleasure to me to know that I have met with some measure of success.

There are many essays in the book that you will not like any more than I do. The secret of this lies in the fact, which you pointed out in your memorable speech at the Linnean Celebration, that no one but a naturalist can really understand Darwin.

I did not go to Cambridge—I had my hands full here. I was not sorry for the excuse. There seemed to me a note of insincerity about the whole business. I am short-tempered. I cannot stand being told that the origin of species has still to be discovered, and that specific differences have no "reality" (Bateson's Essay, p. 89). People are of course at liberty to hold such opinions, but decency might have presented another occasion for ventilating them.—Yours sincerely,

W.T. THISELTON-DYER.

## SIR W.T. THISELTON-DYER TO A.R. WALLACE

*The Ferns, Witcombe, Gloucester. July 11, 1909.*

Dear Mr. Wallace,— ... I have just got F. Darwin's "Foundations." He tries to make out that his father could have dispensed with Malthus. But the selection death-rate in a slightly varying large population is *the* pith of the whole business. The Darwin-Wallace theory is, as you say, "the continuous adjustment of the organic to the inorganic world." It is what mathematicians call "a moving equilibrium." In fact, I have always maintained that it is a mathematical conception.

It seemed to me there was a touch of insincerity about the whole celebration,[38] as the younger Cambridge School as a whole do not even begin to understand the theory.... I take it that the reason is, as you pointed out, that none of them are naturalists.—Yours sincerely,

W.T. THISELTON-DYER.

## TO DR. ARCHDALL REID

*Old Orchard, Broadstone, Dorset. December 28, 1909.*

Dear Dr. Archdall Reid,—Many thanks for your very interesting and complimentary letter. I am very glad to hear of your new book, which I doubt not will be very interesting and instructive. The subjects you treat are, however, so very complex, and require so much accurate knowledge of the facts, and so much sound reasoning upon them, that I cannot possibly undertake the labour and thought required before I should feel justified in expressing an opinion upon your treatment of them....

I rejoice to hear that you have exposed the fallacy of the claims of the Mendelians. I have also tried to do so, but I find it quite impossible for me to follow their detailed studies and arguments. It wants a mathematical mind, which I have not.

But on the general relation of Mendelism to Evolution I have come to a very definite conclusion. This is, that it has no relation whatever to the evolution of species or higher groups, but is really antagonistic to such evolution! The essential basis of evolution, involving as it does the most minute and all-pervading adaptation to the whole environment, is extreme and ever-present plasticity, as a condition of survival and adaptation. But the essence of Mendelian characters is their rigidity. They are transmitted without variation, and therefore, except by the rarest of accidents, can never become adapted to ever-varying conditions. Moreover, when crossed they reproduce the same pair of types in the same proportions as at first, and therefore without selection; they are antagonistic to evolution by continually reproducing injurious or useless characters—which is the reason they are so rarely found in nature, but are mostly artificial breeds or sports. My view is, therefore, that Mendelian characters are of the nature of abnormalities or monstrosities, and that the "Mendelian laws" serve the purpose of eliminating them when, as usually, they are not useful, and thus preventing them from interfering with the normal process of natural selection and adaptation of the more plastic races. I am also glad to hear of your new argument for non-inheritance of acquired characters.—Yours very truly,

ALFRED R. WALLACE.

## TO SIR W.T. THISELTON-DYER

*Old Orchard, Broadstone, Wimborne, February 8, 1911.*

Dear Sir W. Thiselton-Dyer,—I thank you very much for taking so much trouble as you have done in writing your views of my new book.[39] I am glad to find that you agree with much of what I have said in the more evolutionary part of it, and that you differ only on some of my suggested interpretations of the facts. I have always felt the disadvantage I have been under—more especially during the last twenty years—in having not a single good biologist anywhere near me, with whom I could discuss matters of theory or obtain information as to matters of fact. I am therefore the more pleased that you do not seem to have come across any serious misstatements in the botanical portions, as to which I have had to trust entirely to second-hand information, often obtained through a long and varied correspondence.

As to your disagreement from me in the conclusions arrived at and strenuously advocated in the latter portions of my work, I am not surprised. I am afraid, now, that I have not expressed myself sufficiently clearly as to the fundamental phenomena which seem to me absolutely to necessitate a guiding mind and organising power. Hardly one of my critics (I think absolutely not one) has noticed the distinction I have tried and intended to draw between Evolution on the one hand, and the fundamental powers and properties of Life—growth, assimilation, reproduction, heredity, etc.—on the other. In Evolution I recognise the action of Natural Selection as universal and capable of explaining all the facts of the continuous development of species from species, "from am[oe]ba to man." But this, as Darwin, Weismann, Kerner, Lloyd-Morgan, and even Huxley have seen, has nothing whatever to do with the basic mysteries of life—growth, etc. etc. The chemists think they have done wonders when they have produced in their laboratories certain organic substances—always by the use of other organic products—which life builds up within each organism, and from the few simple elements available in air, earth, and water, innumerable structures—bone, horn, hair, skin, blood, muscle, etc. etc.; and these are not amorphous—mere lumps of dead matter—but organised to serve certain definite purposes in each living organism. I have dwelt on this in my chapter on "The Mystery of the Cell." Now I have been unable to find any attempt by any biologist or physiologist to grapple with this problem. One and all, they shirk it, or simply state it to be insoluble. It is here that I state guidance and organising power are essential. My little physiological parable or allegory (p. 296) I think sets forth the difficulty fairly, though by no means adequately, yet not one of about fifty reviews I have read even mentions it.

If you know of any writer of sufficient knowledge and mental power, who has fully recognised and fairly grappled with this fundamental problem, I should be very glad to be referred to him. I have been able to find no approach to it. Yet I am at once howled at, or sneered at, for pointing out the facts that such problems exist, that they are not in any way touched by Evolution, but are far before it, and the forces, laws and agencies involved are those of existences possessed of powers, mental and physical, far beyond those mere mechanical, physical, or chemical forces we see at work in nature....—Yours very truly,

ALFRED R. WALLACE.

## SIR W.T. THISELTON-DYER TO A.R. WALLACE

*The Ferns, Witcombe, Gloucester. February 12, 1911.*

Dear Mr. Wallace,— ... You must let me correct you on one technical point in your letter. It is no longer possible to say that chemists effect the synthesis of organic products "by the use of other organic substances." From what has been already effected, it cannot be doubted that eventually every organic substance will be built up from "the few simple elements available in air, earth and water." I think you may take it from me that this does not admit of dispute....

At any rate we are in agreement as to Natural Selection being capable of explaining evolution "from am[oe]ba to man."

It is generally admitted that that is a mechanical or scientific explanation. That is to say, it invokes nothing but intelligible actions and causes.

De Vries, however, asserts that the Darwinian theory is *not* scientific at all, and that is of course a position he has a right to take up.

But if we admit that it is scientific, then we are precluded from admitting a "directive power."

This was von Baer's position, also that of Kant and of Weismann.

But von Baer remarks that the naturalist is not precluded from asking "whether the *totality* of details leads him to a general and final basis of intentional design." I have no objection to this, and offer it as an olive-branch which you can throw to your howling and sneering critics.

As to "structures organised to serve certain definite purposes," surely they offer no more difficulty as regards "scientific" explanation than the apparatus by which an orchid is fertilised.

We can work back to the am[oe]ba to find ourselves face to face with a scarcely organised mass of protoplasm. And then we find ourselves face to face with a problem which will, perhaps, for ever remain insoluble scientifically. But as for that, so is the primeval material of which it (protoplasm) is composed. "Matter" itself is evaporating, for it is being resolved by physical research into something which is intangible.

We cannot form the slightest idea how protoplasm came into existence. It is impossible to regard it as a mere substance. It is a mechanism. Although the chemist may hope to make eventually all the substances which protoplasm fabricates, and will probably do so, he can only build them up by the most complicated processes. Protoplasm appears to be able to manufacture them straight off in a way of which the chemist cannot form the slightest conception. This is one aspect of the mystery of *life*. Herbert Spencer's definition tells one nothing.

Science can only explain nature as it reveals itself to the senses in terms of consciousness. The explanation may be all wrong in the eyes of omniscience. All one can say is that it is a practical working basis, and is good enough for mundane purposes. But if I am asked if I can solve the riddle of the Universe I can only answer, No. Brunetière then retorts that science is bankrupt. But this is equivocal. It only means that it cannot meet demands beyond its power to satisfy.

I entirely sympathise with anyone who seeks an answer from some other non-scientific source. But I keep scientific explanations and spiritual craving wholly distinct.

The whole point of evolution, as formulated by Lyell and Darwin, is to explain phenomena by known causes. Now, directive power is not a known cause. Determinism compels me to believe that every event is inevitable. If we admit a directive power, the order of nature becomes capricious and unintelligible. Excuse my saying all this. But that is the dilemma as it presents itself to *my* mind. If it does not trouble other people, I can only say, so much the better for them. Briefly, I am afraid I must say that it is ultra-scientific. I think that would have been pretty much Darwin's view.

I do not think that it is quite fair to say that biologists shirk the problem. In my opinion they are not called upon to face it. Bastian, I suppose, believed that he had bridged the gulf between lifeless and living matter. And here is a man, of whom I know nothing, who has apparently got the whole thing cut and dried.—Yours sincerely,

## TO PROF. POULTON

*Old Orchard, Broadstone, Dorset. May 28, 1912.*

My dear Poulton,—Thanks for your paper on Darwin and Bergson.[40] I have read nothing of Bergson's, and although he evidently has much in common with my own views, yet all vague ideas—like "an internal development force"—seem to me of no real value as an explanation of Nature.

I claim to have shown the necessity of an ever-present Mind as the primal cause both of all physical and biological evolution. This Mind works by and through the primal forces of nature—by means of Natural Selection in the world of life; and I do not think I could read a book which rejects this method in favour of a vague "law of sympathy." He might as well reject gravitation, electrical repulsion, etc. etc., as explaining the motions of cosmical bodies....—Yours very truly,

ALFRED R. WALLACE.

## TO MR. BEN R. MILLER

*Old Orchard, Broadstone, Dorset, January 18, 1913.*

Dear Sir,—Thanks for your kind congratulations, and for the small pamphlet[41] you have sent me. I have read it with much interest, as the writer was evidently a man of thought and talent. The first lecture certainly gives an approach to Darwin's theory, perhaps nearer than any other, as he almost implies the "survival of the fittest" as the cause of progressive modification. But his language is imaginative and obscure. He uses "education" apparently in the sense of what we should term "effect of the environment."

The second lecture is even a more exact anticipation of the modern views as to microbes, including their transmission by flies and other insects and the probability that the blood of healthy persons contains a sufficiency of destroyers of the pathogenic germs—such as the white blood-corpuscles—to preserve us in health.

But he is so anti-clerical and anti-Biblical that it is no wonder he could not get a hearing in Boston in 1847.—Yours very truly,

ALFRED R. WALLACE.

## TO PROF. POULTON

*Old Orchard, Broadstone, Dorset. April 2, 1913.*

My dear Poulton,—About two months ago an American ... sent me the enclosed booklet,[42] which he had been told was very rare, and contained an anticipation of Darwinism.

This it certainly does, but the writer was highly imaginative, and, like all the other anticipators of Darwin, did not perceive the whole scope of his idea, being, as he himself says, not sufficiently acquainted with the facts of nature.

His anticipations, however, of diverging lines of descent from a common ancestor, and of the transmission of disease germs by means of insects, are perfectly clear and very striking.

As you yourself made known one of the anticipators of Darwin, whom he himself had overlooked, you are the right person to make this known in any way you think proper. As you have so recently been in America, you might perhaps ascertain from the librarian of the public library in Boston, or from some of your biological friends there, what is known of the writer and of his subsequent history.

If the house at Down is ever dedicated to Darwin's memory it would seem best to preserve this little book there; if not you can dispose of it as you think best.—Yours very truly,

ALFRED R. WALLACE.

P.S.—Two of my books have been translated into Japanese: will you ascertain whether the Bodleian would like to have them?

## TO PROF. POULTON[43]

*Old Orchard, Broadstone, Dorset, June 3, 1913.*

My dear Poulton,—I am very glad you have changed your view about the "Sleeper" lectures being a "fake." The writer was too earnest, and too clear a thinker, to descend to any such trick. And for what? "Agnostic" is not in Shakespeare, but it may well have been used by someone before Huxley. The parts of your Address of which you send me slips are excellent, and I am sure will be of great interest to your audience. I quite agree with your proposal that the "Lectures" shall be given to the Linnean Society.—Yours very truly,

ALFRED R. WALLACE.

## TO MR. E. SMEDLEY

*Old Orchard, Broadstone, Dorset. August 26, 1913.*

Dear Mr. Smedley,—I am glad to see you looking so jolly. I return the photo to give to some other friend. Mr. Marchant, the lecturer you heard, is a great friend of mine, but is now less dogmatic. The Piltdown skull does not prove much, if anything!

The papers are wrong about me. I am not writing anything now; perhaps shall write no more. Too many letters and home business. Too much bothered with many slight ailments, which altogether keep me busy attending to them. I am like Job, who said "the grasshopper was a burthen" to him! I suppose its creaking song.—Yours very truly,

ALFRED R. WALLACE.

## TO MR. W.J. FARMER

*Old Orchard, Broadstone, Wimborne. 1913.*

Dear Sir,— ... I presume your question "Why?" as to the varying colour of individual hairs and feathers, and the regular varying of adjacent hairs, etc., to form the surface pattern, applies to the ultimate cause which enables those patterns to be hereditary, and, in the case of birds, to be reproduced after moulting yearly.

The purpose, or end they serve, I have, I think, sufficiently dealt with in my "Darwinism"; the method by which such useful tints and markings are produced, because useful, is, I think, clearly explained by the law of Natural Selection or Survival of the Fittest, acting through the universal facts of heredity and variation.

But the "why"—which goes further back, to the directing agency which not only brings each special cell of the highly complex structure of a feather into its exactly right position, but, further, carries pigments or produces surface striæ (in the case of the metallic or interference colours) also to their exactly right place, and nowhere else—is the mystery, which, if we knew, we should (as Tennyson said of the flower in the wall) "know what God and Man is."

The idea that "cells" are all conscious beings and go to their right places has been put forward by Butler in his wonderful book "Life and Habit," and now even Haeckel seems to adopt it. All theories of heredity, including Darwin's pangenesis, do not touch it, and it seems to me as fundamental as life and consciousness, and to be absolutely inconceivable by us till we know what life is, what spirit is, and what matter is; and it is probable that we must develop in the spirit world some few thousand million years before we get to this knowledge—if then!

My book, "Man's Place in the Universe," shows, I think, indications of the vast importance of that Universe as the producer of Man which so many scientific men to-day try to belittle, because of what may be, in the infinite!— Yours very truly,

ALFRED R. WALLACE.

# PART IV

## Home Life

(By W.G. WALLACE and VIOLET WALLACE)

In our father's youth and prime he was 6 ft. 1 in. in height, with square though not very broad shoulders. At the time to which our first clear recollections go back he had already acquired a slight stoop due to long hours spent at his desk, and this became more pronounced with advancing age; but he was always tall, spare and very active, and walked with a long easy swinging stride which he retained to the end of his life.

As a boy he does not appear to have been very athletic or muscularly strong, and his shortsightedness probably prevented him from taking part in many of the pastimes of his schoolfellows. He was never a good swimmer, and he used to say that his long legs pulled him down. He was, however, always a good walker and, until quite late in life, capable of taking long country walks, of which he was very fond.

He was very quick and active in his movements at times, and even when 90 years of age would get up on a chair or sofa to reach a book from a high shelf, and move about his study with rapid strides to find some paper to which he wished to refer.

When out of doors he usually carried an umbrella, and in the garden a stick, upon which he leaned rather heavily in his later years. His hair became white rather early in life, but it remained thick and fine to the last, a fact which he attributed to always wearing soft hats. He had full beard and whiskers, which were also white. His eyes were blue and his complexion rather pale. He habitually wore spectacles, and to us he never looked quite natural without them. Towards the end of his life his eyes were subject to inflammation, and the glasses were blue. His hands, though large, were not clumsy, and were capable of very delicate manipulation, as is shown by his skill in handling and preserving insects and bird-skins, and also in sketching, where delicacy of touch was essential. His handwriting is another example of this; it remained clear and even to the end, in spite of the fact that he wrote all his books, articles, and letters with his own hand until the last few years, when he occasionally had assistance with his correspondence; but his last two books, "Social Environment" and "The Revolt of Democracy," written when he was 90 years of age, were penned by himself, and the MSS. are perfectly legible and regular.

He was very domestic, and loved his home. His interest extended to the culinary art, and he was fond of telling us how certain things should be cooked. This became quite a joke among us. He was very independent, and it never seemed to occur to him to ask to have anything done for him if he could do it himself—and he could do many things, such as sewing on buttons and tapes and packing up parcels, with great neatness. When unpacking parcels he never cut the string if it could be untied, and he would fold it up before removing the paper, which in its turn was also neatly folded.

His clothes were always loose and easy-fitting, and generally of some quiet-coloured cloth or tweed. Out of doors he wore a soft black felt hat rather taller than the clerical pattern, and a black overcoat unless the weather was very warm. He wore no ornaments of any kind, and even the silver watch-chain was worn so as to be invisible. He wore low collars with turned-down points and a narrow black tie, which was, however, concealed by his beard. He was not very particular about his personal appearance, except that he always kept his hair and beard well brushed and trimmed.

MRS. A.R. WALLACE (about 1895)

In our early days at Grays we children were allowed to run in and out of his study; but if he was busy writing at the moment we would look at a book until he could give us his attention. His brother in California sent him a live specimen of the lizard called the "horned toad," and this creature was kept in the study, where it was allowed to roam about, its favourite place being on the hearth.

About this time he read "Alice through the Looking-glass," which pleased him greatly; he was never tired of quoting from it and using some of Lewis Carroll's quaint words till it became one of our classics.

Some of our earliest recollections are of the long and interesting walks we took with our father and mother. He never failed to point out anything of interest and tell us what he knew about it, and would answer our numerous questions if possible, or put us off with some joking reference to Boojums or Jabberwocks. We looked upon him as an infallible source of information, not only in our childhood, but to a large extent all his life. When exploring the country he scorned "trespass boards." He read them "Trespassers will be persecuted," and then ignored them, much to our childish trepidation. If he was met by indignant gamekeepers or owners, they were often too much awed by his dignified and commanding appearance to offer any objection to his going where he wished. He was fond of calling our attention to insects and to other objects of natural history, and giving us interesting lessons about them. He delighted in natural scenery, especially distant views, and our walks and excursions were generally taken with some object, such as finding a bee-orchis or a rare plant, or exploring a new part of the country, or finding a waterfall.

In 1876 we went to live at Dorking, but stayed there only a year or two. An instance of his love of mystifying us children may be given. It must have been shortly after our arrival at Dorking that one day, having been out to explore the neighbourhood, he returned about tea-time and said, "Where do you think I have been? To Glory!" Of course we were very properly excited, and plied him with questions, but we got nothing more out of him then. Later on we were taken to see the wonderful place called "Glory Wood"; and it had surely gained in glory by such preparation.

Sometimes it would happen that a scene or object would recall an incident in his tropical wanderings and he would tell us of the sights he had seen. At the time he was greatly interested in botany, in which he was encouraged by our mother, who was an ardent lover of flowers; and to the end of his life he exhibited almost boyish delight when he discovered a rare plant. Many walks and excursions were taken for the purpose of seeing some uncommon plant growing in its natural habitat. When he had found the object of his search we were all called to see it. During his walks and holidays he made constant use of the one-inch Ordnance Maps, which he obtained for each district he visited, planning out our excursions on the map before starting. He had a gift for finding the most beautiful walks by means of it.

In 1878 we moved to Croydon, where we lived about four years. It was at this time that he hoped to get the post of Superintendent of Epping Forest. We still remember all the delights we children were promised if we went to live there. We had a day's excursion to see the Forest, he with his map finding out the roads and stopping every now and then to admire a fresh view or to explain what he would do if the opportunity were given him. It was a very hot day, and we became so thirsty that when we reached a stream, to our great joy and delight he took out of his pocket, not the old leather drinking-cup he usually carried, but a long piece of black indiarubber tubing. We can see him now, quite as pleased as we were with this brilliant idea, letting it down into the stream and then offering us a drink! No water ever tasted so nice! Our mother used to be a little anxious as to the quality of the water, but he always put aside such objections by saying *running* water was quite safe, and somehow we never came to any harm through it. The same happy luck attended our cuts and scratches; he always put "stamp-paper" on them, calling it plaster, and we knew of no other till years later. He used the same thing for his own cuts, etc., to the end of his life, with no ill effects.

In 1881 we moved again, this time to Godalming, where he had built a small house which be called "Nutwood Cottage." After Croydon this was a very welcome change and we all enjoyed the lovely country round. The garden as usual was the chief hobby, and Mr. J.W. Sharpe, our old friend and neighbour in those days, has written his reminiscences of this time which give a very good picture of our father. They are as follows:

> About thirty-five years ago Dr. Wallace built a house upon a plot of ground adjoining that upon which our house stood. I was at that time an assistant master at Charterhouse School; and Dr. Wallace became acquainted with a few of the masters besides myself. With two or three of them he had regular weekly games of chess; for he was then and for long afterwards very fond of that game; and, I understand, possessed considerable skill at it. A considerable portion of his spare time was spent in his garden, in the management of which Mrs. Wallace, who had much knowledge and experience of gardening, very cordially assisted him. Here his characteristic energy and restlessness were conspicuously displayed. He was always designing some new feature, some alteration in a flower-bed, some special environment for a new plant; and always he was confident that the new schemes would be found to have all the perfections which the old ones lacked. From all parts of the world botanists and collectors sent him, from time to time, rare or newly discovered plants, bulbs, roots or seeds, which he, with the help of Mrs. Wallace's practical skill, would try to acclimatise, and to persuade to grow somewhere or other in his garden or conservatory. Nothing disturbed his cheerful confidence in the future, and nothing made him happier than some plan for reforming the house, the garden, the kitchen-boiler, or the universe. And, truth

to say, he displayed great ingenuity in all these enterprises of reformation. Although they were never in effect what they were expected to be by their ingenious author, they were often sufficiently successful; but, successful or not, he was always confident that the next would turn out to be all that he expected of it. With the same confidence he made up his mind upon many a disputable subject; but, be it said, never without a laborious examination of the necessary data, and the acquisition of much knowledge. In argument, of which intellectual exercise he was very fond, he was a formidable antagonist. His power of handling masses of details and facts, of showing their inner meanings and the principles underlying them, and of making them intelligible, was very great; and very few men of his time had it in equal measure.

But the most striking feature in his conversation was his masterly application of general principles: these he handled with extraordinary skill. In any subject with which he was familiar, he would solve, or suggest a plausible solution of, difficulty after difficulty by immediate reference to fundamental principles. This would give to his conclusions an appearance of inevitableness which usually overbore his adversary, and, even if it did not convince him, left him without any effective reply. This, too, had a good deal to do, I am disposed to conjecture, with another very noticeable characteristic of his which often came out in conversation, and that was his apparently unfailing confidence in the goodness of human nature. No man nor woman but he took to be in the main honest and truthful, and no amount of disappointment—not even losses of money and property incurred through this faith in others' virtues—had the effect of altering this mental habit of his.

His intellectual interests were very widely extended, and he once confessed to me that they were agreeably stimulated by novelty and opposition. An uphill fight in an unpopular cause, for preference a thoroughly unpopular one, or any argument in favour of a generally despised thesis, had charms for him that he could not resist. In his later years, especially, the prospect of writing a new book, great or small, upon any one of his favourite subjects always acted upon him like a tonic, as much so as did the project of building a new house and laying out a new garden. And in all this his sunny optimism and his unfailing confidence in his own powers went far towards securing him success.—J.W.S.

"Land Nationalisation" (1882), "Bad Times" (1885), and "Darwinism" (1889) were written at Godalming, also the series of lectures which he gave in America in 1886-7 and at various towns in the British Isles. He also continued to have examination papers[44] to correct each year—and a very strenuous time that was. Our mother used to assist him in this work, and also with the indexes of his books.

We now began to make nature collections, in which he took the keenest interest, many holidays and excursions being arranged to further these engrossing pursuits. One or two incidents occurred at "Nutwood" which have left clear impressions upon our minds. One day one of us brought home a beetle, to the great horror of the servant. Passing at the moment, he picked it up, saying, "Why, it is quite a harmless little creature!" and to demonstrate its inoffensiveness he placed it on the tip of his nose, whereupon it immediately bit him and even drew blood, much to our amusement and his own astonishment. On another occasion he was sitting with a book on the lawn under the oak tree when suddenly a large creature alighted upon his shoulder. Looking round, he saw a fine specimen of the ring-tailed lemur, of whose existence in the neighbourhood he had no knowledge, though it belonged to some neighbours about a quarter of a mile away. It seemed appropriate that the animal should have selected for its attentions the one person in the district who would not be alarmed at the sudden appearance of a strange animal upon his shoulder. Needless to say, it was quite friendly.

A year or so before we left Godalming he enlarged the house and altered the garden. But his health not having been very good, causing him a good deal of trouble with his eyes, and having more or less exhausted the possibilities of the garden, he decided to leave Godalming and find a new house in a milder climate. So in 1889 he finally fixed upon a small house at Parkstone in Dorset.

Planning and constructing houses, gardens, walls, paths, rockeries, etc., were great hobbies of his, and he often spent hours making scale drawings of some new house or of alterations to an existing one, and scheming out the details of construction. At other times he would devise schemes for new rockeries or waterworks, and he would always talk them over with us and tell us of some splendid new idea he had hit upon. As Mr. Sharpe has noted, he was always very optimistic, and if a scheme did not come up to his expectations he was not discouraged but always declared he could do it much better next time and overcome the defects. He was generally in better health and happier when some constructional work was in hand. He built three houses, "The Dell" at Grays, "Nutwood Cottage" at Godalming, and the "Old Orchard" at Broadstone. The last he actually built himself, employing the men and buying all the materials, with the assistance of a young clerk of works; but though the enterprise was a source of great pleasure, it was a constant worry. He also designed and built a concrete garden wall, with which he was very pleased, though it cost considerably more than he anticipated. He had not been at Parkstone long before he set about the planning of "alterations" with his usual enthusiasm. We were both away from home at this time, and consequently had many letters from him, of which one is given as a specimen. His various interests are nearly always referred to in these letters, and in not a few of them his high spirits show themselves in bursts of exuberance which were very characteristic whenever a new scheme was afoot. The springs of eternal youth were for ever bubbling up afresh, so that to us he never grew old. One of us remembers how, when he must have been about 80, someone said, "What a wonderful old man your father is!" This was quite a shock, for to us he was not old. The letter referred to above is the following:

## TO MR. W.G. WALLACE

*Parkstone, Dorset, February 1, 1891.*

My dear Will,—Another week has passed away into eternity, another month has opened its eyes on the world, and still the illustrious Charles [bricklayer] potters about, still the carpenter plies the creaking saw and the stunning hammer, still the plumber plumbs and the bellhanger rattles, still the cisterns overflow and the unfinished drains send forth odorous fumes, still the rains descend and all around the house is a muddle of muck and mire, and still there is so much to do that we look forward to some far distant futurity, when all that we are now suffering will be over, and we may look back upon it as upon some strange yet not altogether uninteresting nightmare!

Briefly to report progress. The new pipe-man has finished the bathroom and nearly done the bells, and we have had gas alight the last three days. The balcony is finished, the bath and lavatory are closed up and waiting for the varnishers. Charles has finished the roof, and the scaffolding is removed. But though two plumbers have tried all their skill, the ball-cock in the cistern won't work, and when the water has been turned on an hour it overflows.

The gutters and pipes to roof are not up, and the night before last a heavy flood of rain washed a quantity of muddy water into the back entrance, which flowed right across the kitchen into the back passage and larder, leaving a deposit of alluvial mud that would have charmed a geologist. However, we have stopped that for the future by a drain under the doorstep. The new breakfast-room is being papered and will look tidy soon. A man has been to measure for the stairs. The front porch door is promised for to-morrow, and the stairs, I suppose, in another week. A lot of fresh pointing is to be done, and all the rain-water pipes and the rain-water cistern with its overflow pipes, and then the greenhouse, and then all the outside painting—after which we shall rest for a month and then do the inside papering; but whether that can be done before Easter seems very doubtful....

Our alterations still go on. The stairs just up—Friday night we had to go outside to get to bed, and Saturday and Sunday we *could* get up, but over a chasm, and with alarming creaks. Now it is all firm, but no handrail yet. Painters still at work, and whitewashers. Porch door up, with two birds in stained glass—looks fine—proposed new name, "Dicky-bird Lodge." Bath fixed, but waiting to be varnished—luxurious!...

Dr. Wallace had already received four medals from various scientific societies, and at our suggestion he had a case made to hold them all, which is referred to in the following letter. The two new medals mentioned were those of the Royal Geographical and Linnean Societies. He attached very little importance to honours conferred upon himself, except in so far as they showed acceptance of "the truth," as he called it.

## TO MISS VIOLET WALLACE

*Parkstone, Dorset. April 3, 1892.*

My dear Violet,— ... I have got J.G. Wood's book on the horse. It is very good; I think the best book he has written, as his heart was evidently in it....

A dreadful thing has happened! Just as I have had my medal-case made, "regardless of expense," they are going to give me another medal! Hadn't I better decline it, with thanks? "No room for more medals"!!—Your affectionate papa,

ALFRED R. WALLACE.

P.S.—A poor man came here last night (Saturday) with a basket of primrose roots—had carried them eight miles, couldn't sell one in Poole or Parkstone—was 64 years old—couldn't get any work to do—had no home, etc. So, though I do not approve of digging up primrose roots as a trade, I gave him 1s. 6d. for them, pitying him as one of the countless victims of landlordism.—A.R.W.

A poor man was sentenced to fourteen days' hard labour last week for picking snowdrops in Charborough Park. Shame!—A.R.W., Pres. L.N. Society.

### TO Miss VIOLET WALLACE

*Parkstone, Dorset. May 5, 1892.*

My dear Violet,—I have finished reading "Freeland." It is very good—as good a story as "Looking Backward," but not quite so pleasantly written—rather heavy and Germanic in places. The results are much the same as in "Looking Backward" but brought about in a different and very ingenious manner. It may be called "Individualistic Socialism." I shall be up in London soon, I expect, to the first Meetings of the Examiners in the great science of "omnium gatherum."[45]—Your affec. papa,

ALFRED R. WALLACE.

While he lived at Parkstone our father built a small orchid house in which he cultivated a number of orchids for a few years, but the constant attention which they demanded, together with the heated atmosphere, were too much for him, and he was obliged to give them up. He was never tired of admiring their varied forms and colours, or explaining to friends the wonderful apparatus by which many of them were fertilised. The following letter shows his enthusiasm for orchids:

### TO Miss VIOLET WALLACE

*Parkstone, Dorset. November 25, 1894.*

My dear Violet,— ... I have found a doctor at Poole (Mr. Turner) who has two nice orchid houses which he attends to entirely himself, and as I can thus get advice and sympathy from a fellow maniac (though he *is* a public vaccinator!) my love of orchids is again aroused to fever-heat, and I have made some alterations in the greenhouse which will better adapt it for orchid growing, and have bought a few handsome kinds very cheap, and these give me a lot of extra work and amusement....

## TO HIS WIFE

*Hôtel du Glacier du Rhône. Wednesday evening, [July, 1895].*

My dear Annie,—I send you now a box of plants I got on both sides of the Furka Pass yesterday, and about here to-day. The Furka Pass on both sides is a perfect flower-garden, and the two sides have mostly different species. The violets and anemones were lovely, and I have got two species of glorious gentians.... All the flowers in the box are very choice species, and have been carefully dug up, and having seen how they grow, I have been thinking of a plan of making a little bed for them on the top of the new rockery where there is now nothing particular. Will you please plant them out carefully in the zinc tray of peat and sphagnum that stands outside near the little greenhouse door? Just lift up the sphagnum and see if the earth beneath is moist, if not give it a soaking. Then put them all in, the short-rooted ones in the sphagnum only, the others through into the peat. Then give them a good syringing and put the tray under the shelf outside the greenhouse, and cover with newspaper for a day or two. After that I think they will do, keeping them moist if the weather is dry. I am getting hosts of curiosities. To-day we found four or five species of willows from 1/4 in. to 2 in. high, and other rarities.... In haste for post and dinner.—Your ever affectionate

ALFRED R. WALLACE.

## TO Miss VIOLET WALLACE

*Parkstone, Dorset. October 22, 1897.*

My dear Violet,—In your previous letter you asked me the conundrum, Why does a wagtail wag its tail? That's quite easy, on Darwinian principles. Many birds wag their tails. Some Eastern flycatchers—also black and white—wag their long tails up and down when they alight on the ground or on a branch. Other birds with long tails jerk them up in the air when they alight on a branch. Now these varied motions, like the motions of many butterflies, caterpillars, and many other animals, must have a use to the animal, and the most common, or rather the most probable, use is, either to frighten or to distract an enemy. If a hawk was very hungry and darted down on a wagtail from up in the air, the wagging tail would be seen most distinctly and be aimed at, and thus the bird would be missed or at most a feather torn out of the tail. The bird hunts for food in the open, on the edges of ponds and streams, and would be especially easy to capture, hence the wagging tail has been developed to baffle the enemy....

## TO Miss VIOLET WALLACE

*Parkstone, Dorset. March 8, 1899.*

My dear Violet,— ... I have now finished reading the "Maha Bharata," which is on the whole very fine—finer, I think, than the "Iliad." I have read a good deal of it twice, and it will bear reading many times. It corresponds pretty nearly in date with the "Iliad," the scenes it describes being supposed to be about B.C. 1500. Many of the ideas and moral teachings are beautiful; equal to the best teaching and superior to the general practice of to-day. I have made a lot of emendations and suggestions, which I am going to send to the translator, as the proofs have evidently not been carefully read by any English literary man.

About the year 1899 Dr. Wallace began to think of leaving Parkstone, partly for reasons of health and partly to get a larger garden, if possible. He spent three years in looking for a suitable spot in many of the southern counties, and we were all pressed to join in the search. Finally he found just the spot he wanted at Broadstone; only three miles away. The following letters describe his final success—all written with his usual optimism and high spirits:

## TO MR. W.G. WALLACE

*Parkstone, Dorset. October 26, 1901.*

My dear Will,—At length the long quest has come to an end, and I have agreed to buy three acres of land at Broadstone. Ma and I have just been over again this morning to consider its capabilities, and the exact boundaries that will be the most advantageous, as I have here the great advantage of choosing exactly what I will have. I only wish I could afford five acres instead of three, or even ten; but the three will contain the very eye of the whole. I enclose you a bit of the 6-inch ordnance on which I have marked the piece I have finally fixed upon in red chalk. The attractive bit is the small enclosure of one acre, left rather paler, which is an old orchard in a little valley sloping downward to the S.S.E. There are, perhaps, a score of trees in it—apples, pears, plums and cherries, I believe, and under them a beautiful green short turf like a lawn—kept so, I believe, by rabbits. From the top of this orchard is a fine view over moor and heather, then over the great northern bay of Poole Harbour, and beyond to the Purbeck Hills and out to the sea and the Old Harry headland. It is not very high—about 140 feet, I think, but being on the edge of one of the plateaus the view is very effective. On the top to the left of the road track is a slightly undulating grass field, of which I have a little less than an acre. To the right of the fence, and coming down to the wood, is very rough ground densely covered with heather and dwarf gorse, a great contrast to the field. The wood on the right is mixed but chiefly oak, I think, with some large firs, one quite grand; while the wood on the left is quite different, having some very tall Spanish chestnuts loaded with fruit, some beeches, some firs—but I have not had time yet to investigate thoroughly. Thus this little bit of three acres has five subdivisions, each with a quite distinct character of its own, and I never remember seeing such variety in such a small area. The red wavy line is about where I shall have to make my road, for the place has now no road, and I think I am very lucky in discovering it and in getting it. Another advantage is in the land, which is varied to suit all crops. I fancy ... I shall find places to grow most of my choice shrubs, etc., better than here. I expect bulbs of all kinds will grow well, and I mean to plant a thousand or so of snowdrops, crocuses, squills, daffodils, etc., in the orchard, where they will look lovely.

### TO MR. W.G. WALLACE

*Parkstone, Dorset. November 6, 1901.*

My dear Will,— ... I have taken advantage of a foggy cold day to trace you a copy of the ground plan of the proposed house.... Of course the house will be much larger than we want, but I look to future value, and rather than build it smaller, to be enlarged afterwards, I would prefer to leave the drawing-room and bedroom adjoining with bare walls inside till they can be properly finished. The house-keeper's room would be a nice dining-room, and the hall a parlour and drawing-room combined. But the outside must be finished, on account of the garden, creepers, etc. The S.E. side (really about S.S.E.) has the fine views. If you can arrange to come at Christmas we will have a picnic on the ground the first sunny day. I was all last week surveying—a very difficult job, to mark out exactly three acres so as to take in exactly as much of each kind of ground as I wanted, and with no uninterrupted view over any one of the boundary lines! I found the sextant, and it was very useful setting out the two right angles of the northern boundary. I have not got possession yet, but hope to do so by next week. The house, we reckon, can be built for £1,000 at the outside....

### TO MRS. FISHER

*Parkstone, Dorset. February 4, 1902.*

Dear Mrs. Fisher,— ... You will be surprised to hear that I have been so rash as to buy land and to (propose to) build a house! Every other effort to get a pleasant country cottage with a little land having failed, we discovered, accidentally, a charming spot only four miles from this house and half a mile from Broadstone Station, and have succeeded in buying three acres, *chosen by myself*, from Lord Wimborne at what is really a reasonable price. In its contour, views, wood, and general aspect of wild nature it is almost perfection; and Annie, Violet, and Will are all pleased and satisfied with it. It is on the slope of the Broadstone middle plateau, looking south over Poole Harbour with the Purbeck Hills beyond, and a little eastward out to the sea.... The ground is good loam in the orchard, with some sand and clay in the field, but this is so open to the sun and air that we are not afraid of it, as the *house-site* will be entirely concreted over, and I have arranged for a heating stove in a cellar, which will warm and dry the whole basement. In a week or two we hope to begin building, so you may fancy how busy I am, especially as we are building it without a contractor, with the help of a friend.... I go over two or three times a week, as I have two gardeners at work. In the summer (should I be still in the land of the living) I hope you will be able to come and see our little estate, which is to be called by the descriptive name of "Old Orchard." I have got a good architect to make the working drawings and he has designed a very picturesque yet unpretentious house.—Yours very truly,

ALFRED R. WALLACE.

## TO MR. W.G. WALLACE

*Parkstone, Dorset. March 2, 1902.*

My dear Will,—This week's progress has been fairly good although the wet after the frost has caused two falls in the cellar excavations, and we have had to put drain pipes to carry water out, though not much accumulated.... During the week some horses in the field have not only eaten off the tops of the privet hedge, but have torn up some dozens of the plants by the roots, by putting their heads over the 4-foot wire fence. I am therefore obliged in self-defence to raise the post a foot higher and put barbed wire along the top of it. Some cows also got in our ground one day and ate off the tops of the newly planted laurels, which I am told they are very fond of, so I have got a chain and padlock for our gate....

We moved into the new house at Broadstone at the end of November, 1902, before it was quite finished, and here Dr. Wallace lived till the end of his life. The garden was an endless source of interest and occupation, being much larger than any he had had since leaving Grays.

When writing he was not easily disturbed and never showed any impatience or annoyance at any interruption. If interrupted by a question he would pause, pen in hand, and reply or discuss the matter and then resume his unfinished sentence.

THE STUDY AT "OLD ORCHARD"

He seemed to have the substance of his writing in his mind before he commenced, and did not often refer to books or to notes, though he usually had one or two books or papers on the table at hand, and sometimes he would jump up to get a book from the shelves to verify some fact or figure. When preparing for a new book or article he read a great many works and papers bearing on the subject. These were marked with notes and references on the flyleaves; and often by pencil marks to indicate important passages, but he did not often make separate notes. He had a wonderful memory, and stored in his mind the facts and arguments he wished to use, or the places where they were to be found. He borrowed many books from libraries, and from these he sometimes made a few notes. He was not a sound sleeper, and frequently lay awake during the night, and then it was that he thought out and planned his work. He often told us with keen delight of some new idea or fresh argument which had occurred to him during these waking hours.

After spending months, or sometimes years, in reading and digesting all the literary matter he could obtain on a subject,—and forming a plan for the treatment of it, he would commence writing, and keep on steadily for five or six hours a day if his health permitted. He also wrote to people all over the world to obtain the latest facts bearing on the subject.

In 1903 he began writing "Man's Place in the Universe."

## TO MR. W.G. WALLACE

*Old Orchard. July 8, 1903.*

My dear Will,—I have just finished going over your notes and corrections of the last four chapters. I can't think how I was so stupid to make the mistake in figures which you corrected. In almost all cases I have made some modification in accordance with your suggestions, and the book will be much improved thereby. I have put in a new paragraph about the stars in other parts than the Milky Way and Solar Cluster, but there is really nothing known about them. I have also cut out the first reference to Jupiter altogether. Of course a great deal is speculative, but any reply to it is equally speculative. The question is, which speculation is most in accordance with the known facts, and not with prepossessions only?

Considering that the book has all been read up and written in less than three months, it cannot be expected to be as complete and careful as if three years had been expended on it, but then it is fresher perhaps. The bit about the pure air came to me while writing, and I let myself go. Why should I not try and do a little good and make people think a little on such matters, when I have the chance of perhaps more readers than all my other books?

As to my making too much of Man, of course that is the whole subject of the book! And I look at it differently from you, because I know *facts* about him you neither know nor believe *yet*. If you are once convinced of the facts and teachings of Spiritualism, you will think more as I do.

The following letter refers to his little book on Mars.

## TO MR. W.G. WALLACE

*Broadstone, Wimborne. September 26, 1907.*

My dear Will,— ... After elaborate revision and correction I have sent my MS. of the little "Mars" book to Macmillans yesterday.... Will you read the whole proofs carefully, in the character of the "intelligent reader"? Your fresh eye will detect little slips, bad logic, too positive statements, etc., which I may have overlooked. It will only be about 100 or 150 pages large type—and I want it to be really good, and free from blunders that any fool can see....

For some years now he had suffered from repeated attacks of asthma and bronchitis. He had tried the usual remedies for these complaints without any good results, and, though still able to write, had then no thought of beginning any large work; in fact, he considered he had but a few more years to live. When Mr. Bruce-Joy came to see him in order to model the portrait medallion, he mentioned in the course of conversation that he had tried the Salisbury treatment with wonderful results. Our father was at first incredulous, but decided to try it in a modified form. He gave up all starchy foods and ate beef only, cooked in a special manner to render it more digestible. He found such relief from this change of diet that from this time onwards he followed a very strict daily routine, which he continued to the end of his life with slight variations.

He made himself a cup of tea on a gas stove in his bedroom at 6 a.m. (the exact quantity of tea and water having been measured the previous evening), and boiled it in a small double saucepan for a definite time by the watch. He always said this cup of tea tasted better than at any other time of the day. He then returned to bed and slept till 8 a.m. During his last two or three years he suffered from rheumatism in his shoulder and it took him a long time to dress, and he called in the aid of his gardener in the last year, who acted as his valet. While dressing he prepared a cup of cocoa on the gas stove, which he carried into the study (next door) at 9 a.m. This was all he had for breakfast, and he took it while reading the paper or his letters.

Dinner at one o'clock was taken with his family, and he usually related any interesting or striking news he had read in the paper, or in his correspondence, and commented upon it, or perhaps he would tell us of some new flower in the garden.

He drank hot water with a little Canary sack and a dash of soda-water, to which he added a spoonful of plum jam. He was very fond of sweet things, such as puddings, but he had to partake sparingly of them, and it was a great temptation when some dish of which he was particularly fond was placed upon the table.

After dinner he usually took a nap in the study before resuming work or going into the garden.

Tea was at four o'clock, and consisted only of a cup of tea, which he made himself in the study, unless there were visitors whom he wished to see, when he would sometimes take it into the drawing-room and make it there.

After tea he again wrote, or took a turn in the garden if the weather and season permitted. Latterly he spent a good part of the afternoon and evening reading and dozing on the sofa, and only worked at short intervals when he felt equal to it.

Supper, at seven, was a repetition of dinner, and he took it with us in the dining-room. After supper he generally read a novel before the fire except in the very hottest weather, and he frequently dozed on and off till he retired at eleven. He made himself a cup of cocoa while preparing for bed, and drank it just before lying down.

For the last year or two it was a constant difficulty with him to secure enough nourishment without aggravating his ailments by indigestion. During this time he suffered continuous discomfort, though he seldom gave utterance to complaint or allowed it to affect the uniform equability of his temper.

In 1903 his daughter came to live with her parents, who generously allowed her to take three or four children as pupils. At first we feared they might bother our father, but he really enjoyed seeing them about and talking to them. He was always interested in any new child, and if for a short time none were forthcoming, always lamented the fact. At dinner the children would ask him all sorts of questions, very amusing ones sometimes. They were also intensely interested in what he ate, and watched with speechless wonder when they saw him eating orange, banana, and sugar with his meat.

One of these early pupils, Reginald B. Rathbone, has sent reminiscences which are so characteristic that we give them as they stand:

"I have stayed at Dr. Wallace's house on three occasions; the first two were when I was only about eight or nine years old, and my recollections of him at that time are therefore necessarily somewhat dim. Certain things, however, have stuck in my memory. I went there quite prepared to see a very venerable and imposing-looking old gentleman, and filled in advance with much awe and respect for him. As regards his personal appearance I was by no mean disappointed, as his tall, slightly-stooping figure, long white hair and beard, and his spectacles fulfilled my highest expectations, I remember being struck with the kindly look of his eyes, and indeed they did not belie his nature, for he always treated me with great kindness, patience and indulgence, which is somewhat remarkable considering my age, and how exasperating I must have been sometimes. I soon began to regard him as a never-failing fount of wisdom, and as one who could answer any question one liked to put to him. Of this latter fact I was not slow to take advantage. I plied him with every kind of question my imaginative young brain could conceive, usually beginning with 'why.'

"He nearly always gave me an answer, and what is more, a satisfactory one, and well within the scope of my limited understanding. These definite, satisfactory answers of his used to afford me great pleasure, it being quite a new experience for me to have all my questions answered for me in this way. These answers, as I have said, were nearly always forthcoming, though indeed, on one or two occasions, in answer to an especially ridiculous query of mine he would answer, 'That is a very foolish question, Reggie.' But this was very rare.

"I remember taking a great interest in what Dr. Wallace ate. He had a hearty appetite, and was no believer in vegetarianism, for at lunch his diet consisted chiefly of cold beef, liberally seasoned with various sauces and relishes, also vinegar. I used to gaze at these bottles with great admiration. Whenever there were peas he used to take large quantities of sugar with them. This greatly aroused my curiosity, and I questioned him about it. 'Why,' said he, 'peas themselves contain sugar; it is, therefore, much more sensible to take sugar with them than salt.' And he recounted an anecdote of how an eminent personage he had once dined with had been waited on with great respect and attention by all present, but salt was offered to him with the peas. 'If you want to make me quite happy,' said the great man, 'you will give me some sugar with my peas.' His favourite drink, I remember, was Canary sack.

"He had a strongly humorous side, and always enjoyed a good laugh. As an instance of this, I will recount the following incident: When I had returned home after my first visit to 'The Old Orchard,' my sister, three years older than myself, and I had a heated argument on the subject of the number of stomachs in a cow. I insisted it was three; she, on the other hand, held that it was seven. After a long and fierce dispute, I exclaimed: 'Well, let us write to Dr. Wallace, and he will settle it for us and tell us the real number.' This we did, the brazen audacity of the proceeding not striking us at the time. By return of post we received a letter which, alas! I have unfortunately not preserved, but the substance of which I well remember. 'Dear Irene and Reggie,' it ran, 'Your dispute as to the number of stomachs which a cow possesses can be settled and rectified by a simple mathematical process usually called subtraction, thus:

| Irene's Cow | 7 stomachs |
| Reggie's Cow | 3 stomachs |
| The Farmer's cow | 4 stomachs |

"Dr. Wallace then went on to explain the names and uses of the four stomachs.

"Two instances of his fun come to my mind as I write. 'Why,' I asked, 'do you sometimes take off your spectacles to read the paper?' 'Because I can see better without 'em,' he said. 'Then why,' I asked again, 'do you ever wear them?' 'Because I can see better with 'em,' was the reply. The other instance relates to chloroform. He was describing the agonies suffered by those who had to undergo amputation before the discovery of anæsthetics, whereas nowadays, he said, 'you are put under chloroform, then wake up and find your arm cut off, having felt nothing. Or you wake up and find your leg cut off. Or you wake up and find your head cut off!' He then laughed heartily at his own joke.

"These are just a few miscellaneous reminiscences, many of them no doubt trivial, but they may perhaps be not entirely devoid of interest, when it is remembered that they are the impressions and recollections of one who was then a boy of eight years old."—B.B.K.

The year 1908 was very auspicious to Dr. Wallace. To begin with, it was the fiftieth anniversary of the reading of the Darwin and Wallace joint papers on the Origin of Species before the Linnean Society, an event which was commemorated in the way described elsewhere.

In the autumn, and just as he was beginning to recover from a spell of bad health, he was invited to give a lecture at the Royal Institution, the prospect of which seemed to have upon him a most stimulating effect; he at once began to think about a suitable subject.

Following closely on this came the news that the Order of Merit was to be conferred upon him. His letters to his son give the details of this eventful period:[46]

### TO MR. W.G. WALLACE

*Old Orchard, Broadstone, Wimborne. October* 28, 1908.

My dear Will,— ... I have a rather surprising bit of news for you. When I was almost at my worst, feeling very bad, I had a letter inviting me to give an evening lecture at the Royal Institution, for their Jubilee of the "Origin of Species"! Of course I decided at once to decline as impossible, etc., having nothing new to say, etc. But a few hours afterwards an idea suddenly came to me for a very fine lecture, if I can work it out as I hope—and the more I thought over it the better it seemed. So, two days back, I wrote to Sir W. Crookes—the Honorary Secretary, who had written to me—accepting provisionally!... Here is another "crowning honour"—the most unexpected of all!...

### TO MR. W.G. WALLACE

*Old Orchard, Broadstone, Wimborne. December 2, 1908.*

My dear Will,— ... This morning the Copley Medals came, gold and silver, smaller than any of the others, but very beautifully designed; the face has the Royal Society's arms, with Copley's name, and "Dignissimo," and my name below. The reverse is the Royal Arms. By the same post came a letter from the Lord Chancellor's Office informing me, to my great relief, that the King had been graciously pleased to dispense with my personal attendance at the investiture of the Order of Merit, ...

### TO MR. W.G. WALLACE

*Old Orchard, Broadstone, Wimborne. December 17, 1908.*

My dear Will,—The ceremony is over, very comfortably. I am duly "invested," and have got two engrossed documents, both signed by the King, one appointing me a member of the "Order of Merit" with all sorts of official and legal phrases, the other a dispensation from being personally "invested" by the King—as Col. Legge explained, to safeguard me as having a right to the Order in case anybody says I was not "invested." ... Colonel Legge was a very pleasant, jolly kind of man, and he told us he was in attendance on the German Emperor when he was staying near Christchurch last summer, and went for many drives with the Emperor only, all about the country.... Col. Legge got here at 2.40, and had to leave at 3.20 (at station), so we got a carriage from Wimborne to meet the train and take him back, and Ma gave him some tea, and he said he had got a nice little place at Stoke Poges but with no view like ours, and he showed me how to wear the Order and was very pleasant: and we were all pleased....

The next letter refers to the discovery of a rare moth and some beetles in the root of an orchid. It was certainly a strange yet pleasant coincidence that these creatures should find themselves in Dr. Wallace's greenhouse, where alone they would be noticed and appreciated as something uncommon.

### TO MR. W.G. WALLACE

*Old Orchard, Broadstone, Wimborne. February 23, 1909.*

My dear Will,— ... In my last letter I did not say anything about my morning at the Nat. Hist. Museum.... What I enjoyed most was seeing some splendid New Guinea butterflies which Mr. Rothschild[17] and his curator, Mr. Jordan, brought up from Tring on purpose to show me. I could hardly have imagined anything so splendid as some of these. I also saw some of the new paradise birds in the British Museum. But Mr. Rothschild says they have five times as many at Tring, and much finer specimens, and he invited me to spend a week-end at Tring and see the Museum. So I may go, perhaps—in the summer.

But I have a curious thing to tell you about insect collecting at "Old Orchard." About five months back I was examining one of the clumps of an orchid in the glass case—which had been sent me from Buenos Ayres by Mr. John Hall—when three pretty little beetles dropped out of it, on the edge of the tank, and I only managed to catch two of them. They were pretty little Longicornes, about an inch long, but very slender and graceful, though only of a yellowish-brown colour. I sent them up to the British Museum asking the name, and telling them they could keep them if of any use. They told me they were a species of the large South American genus Ibidion, but they had not got it in the collection!

On the Sunday before Christmas Day I was taking my evening inspection of the orchids, etc., in the glass case when a largish insect flew by my face, and when it settled it looked like a handsome moth or butterfly. It was brilliant orange on the lower wings, the upper being shaded orange brown, very moth-like, but the antennæ were clubbed like a butterfly's. At first I thought it was a butterfly that mimicked a moth, but I had never seen anything like it before.

Next morning I got a glass jar half filled with bruised laurel leaves, and Ma got it in, and after a day or two I set it, clumsily, and meant to take it to London, but had no small box to put it in. I told Mr. Rothschild about it, and he said it sounded like a Castnia—curious South American moths very near to butterflies. So he got out the drawer with them, but mine was not there; then he got another drawer half-empty, and there it was—only a coloured drawing, but exactly like. It had been described, but neither the Museum nor Mr. Rothschild had got it! I had had the orchids nearly a year and a half, so it must have been, in the chrysalis all that time and longer, which Mr. Rothschild said was the case with the Castnias. On going home I searched, and found the brown chrysalis-case it had come out of among the roots of the same orchid the little Longicornes had dropped from. It is, I am pretty sure, a Brazilian species, and I have written to ask Mr. Hall if he knows where it came from. I have sent the moth and chrysalis to Prof. Poulton (I had promised it to him at the lecture) for the Oxford collection, and he is greatly pleased with it; and especially with its history—one quite small bit of an orchid, after more than a year in a greenhouse, producing a rare or new beetle and an equally rare moth!...

I am glad to say I feel really better than any time the last ten years.—A.R.W.

The Rev. O. Pickard-Cambridge has kindly written his reminiscence of another very curious coincidence connected with a natural history object.

"Some years ago, on looking over some insect drawers in my collection, Mr. A.R. Wallace exclaimed, 'Why, there is my old Sarawak spider!' 'Well! that is curious,' I replied, 'because that spider has caused me much trouble and thought as to who might have caught it, and where; I had only lately decided to describe and figure it, even though I could give the name of neither locality nor finder, being, as it seemed to me, of a genus and species not as yet recorded; also I had, as you see, provisionally conferred your name upon it, although I had not the remotest idea that it had anything else to do with you.' 'Well,' said Mr. Wallace, 'if it is my old spider it ought to have my own private ticket on the pin underneath.' 'It has a ticket,' I replied, 'but it is unintelligible to me; the spider came to me among some other items by purchase at the sale of Mr. Wilson Saunders' collections.' 'If it is mine,' said Wallace (examining it), 'the ticket should be so-and-so. And it is! I caught this spider at Sarawak, and specially noted its remarkable form. I remember it as if it were yesterday, and now I find it here, and you about to publish it as a new genus and species to which, in total ignorance of whence it came or who caught it, you have given my name!' Thus it stands, and '*Friula Wallacii*, Camb. (family Gasteracanthidæ), taken by Alfred Russel Wallace at Sarawak,' is the (unique as I believe) type specimen, in my collection."—O.P.C.

Dr. Wallace was very fond of reading good novels, and usually spent an hour or two, before retiring to bed, with what he called a "good domestic story." One of his favourite authors was Marion Crawford. Poetry appealed to him very strongly, and he had a good memory for his favourite verses, especially for those he had learned in his youth. Amongst his books were over fifty volumes of poetry.

He liked to see friends or interesting visitors, but he was rather nervous with strangers until he became interested in what they had to say. He enjoyed witty conversation, and especially a good story well told. No one laughed more heartily than he when he was much amused, and he would slap his hands upon his knees with delight.

He was very accessible to anyone who might have something to say worth hearing, and he had a great many visitors, especially during the last ten years of his life. Many people distinguished in science, literature, or politics called upon him, and he always enjoyed these visits, and the excitement of them seemed to have no bad effect upon him, even in the last year, when we sometimes feared he might be fatigued by them. In consequence of his sympathy with many heterodox ideas he frequently had visits from "cranks" who wished to secure his support for some new theory or "discovery." He would listen patiently, perhaps ask a few questions, and then endeavour to point out their fallacies. He would amuse us afterwards by describing their "preposterous ideas," and if much bored, he would speak of them as "muffs." He was loath to hurt their feelings, but he generally ended by expressing his opinion quite clearly, occasionally to their discomfiture.

Dr. Littledale has contributed some reminiscences which may be introduced here.

"When I first met Dr. Wallace the conversation turned on the types of visitors that came to see him, and he gave us an amusing account of two young women who called on him to read through a most ponderous treatise relating to the Universe (I think it was). At all events the treatise proved, amongst other things, that Kepler's laws were all wrong. Dr. Wallace was very busy at the time, and politely declined to undertake the task. I remember him well describing with his hands the size of this enormous manuscript and laughing heartily as he detailed how the writer of the manuscript, the elder of the two sisters, persistently tried to persuade him that her theories were all absolutely proved in the work, while the younger sister acted as a sort of echo to her sister. The climax came in a fit of weeping, and, as Dr. Wallace described it, the whole fabric of the universe was washed away in a flood of tears.

"On one occasion, when I was asked by Mrs. Wallace to see Dr. Wallace professionally, he was lying on the sofa in his study by the fire wrapped up in rugs, having just got over a bad shivering attack or rigor. His temperature was 104° Fahr., and all the other usual signs of acute fever were present, but nothing to enable one to form a positive opinion as to the cause. It must have been forty years since he had been in the tropics, but I think he felt that it was an attack of malarial fever. Knowing my patient, my treatment consisted in asking what he was going to do for himself. 'Well,' he said, 'I am going to have a hot bath and then go to bed, and to-morrow I shall get up and go into the garden as usual.' And he was out in the garden next day when I went to see him. This was an instance, doubtless one of many, of the 'will to live,' which carried him through a long life.

"Once, when he was talking about the gaps in the evolution of life, viz. between the inorganic and organic, between vegetable and animal, and between animal and man, I asked, 'Why postulate a beginning at all? We are satisfied with illimitability at one end, why not at the other?' 'For the simple reason,' he said, 'that the mind cannot comprehend anything that has never had a beginning.'

"What attracted me to him most, I think, was his remarkable simplicity of language, whatever the topic of conversation might be, and this not the simplicity of the great mind bringing itself down to the level of the ordinary individual, but his customary mode of expression. I have heard him say that he felt the need of the fluency of speech which Huxley possessed, as he had to cast about for the expression that he wanted. This may have been the case when he was lecturing, but I certainly never noticed it in conversation."—H.E.L.

Dr. Wallace was always interested in young men and others who were going abroad with the intention of studying Natural History, and gave them what advice and help he could. He much enjoyed listening to the accounts given by travellers of the scenes, animals and plants and native life they had seen, and deplored the so-called civilising of the natives, which, in his opinion, generally meant their exploitation by Europeans, leading to their deterioration and extermination.

His nervousness with strangers sometimes led them to form quite erroneous impressions. It occasionally found expression in a nervous laugh which had nothing to do with amusement or humour, but was often heard when he was most serious and felt most deeply. One or two interviewers described it as a "chuckle," an expression which suggested feelings most opposite to those which he really experienced.

Although he could draw and sketch well, he did not take much pleasure in it, and only exercised his skill when there was a definite object in view. His sketches show a very delicate touch, and denote painstaking accuracy, while some are quite artistic. He much preferred drawing with compasses and squares, there being a practical object in his mind for which the plans or drawings were only the first steps. Even in his ninety-first year he found much enjoyment in drawing plans, and spent many hours in designing alterations to a small cottage which his daughter had bought.

He was interested in literary puzzles and humorous stories, and he preserved in an old scrap-book any that appealed to him. He would sometimes read some of them on festive occasions, or when we had children's parties, and sometimes he laughed so heartily himself that he could not go on reading.

In reviewing the years during which Dr. Wallace lived at Broadstone, the last decade, when he was between eighty and ninety years of age, this period seems to have been one of the most eventful, and as full of work and mental activity as any previous period. He never tired of his garden, in which he succeeded in growing a number of rare and curious shrubs and plants. Our mother shared his delight and interest in the garden, and knew a great deal about flowers. She had an excellent memory for their botanical names, and he often asked her the name of some plant which he was pointing out to a friend and which for the moment he had forgotten. She was very fond of roses and of primroses, and there was a fine display of these flowers at "Old Orchard." She was successful in "budding" and in hybridising roses, and produced several beautiful varieties. She was proficient in raising seeds, and he sometimes placed some which he received from abroad in her charge.

When he first came to live at Broadstone he frequently took short walks to the post or to the bank, and sometimes went by train to Poole on business, but he gradually went out less and less, till in the last few years he seldom went outside the garden, but strolled about looking at the flowers or supervising the construction of a new bed or rockery. During his last years his gardener wheeled him about the garden in a bath-chair when he did not feel strong enough to walk all the time.

In 1913, after his last two small books were written, he did no more writing except correspondence. This he attended to himself, except on one or two occasions when he was not very well or felt tired, when he asked one of us to answer a few letters for him. He took great interest in a small cottage which had recently been acquired on the Purbeck Hills near the sea, and in September, much against our wishes, he went there for two nights, taking the gardener to look after him. Luckily the weather was fine, and the change and excitement seemed to do him good, and during the next month he was very bright and cheerful, though, as some of his letters to his old friend Dr. Richard Norris and to Dr. Littledale show, he had been becoming increasingly weak.

### TO MISS NORRIS

*Old Orchard, Broadstone, Dorset. December 10, 1912.*

My dear Miss Norris,—I am very sorry to hear that your father is so poorly. The weather is terribly gloomy, and I have not been outside my rooms and greenhouse for more than an hour a week perhaps, for the last two months, and feel the better for it. Just now I feel better than I have done for a year past, having at last, I think, hit upon a proper diet, though I find it very difficult to avoid eating or drinking too much of what I like best.... It is one of my fads that I hate to waste anything, and it is that partly which makes it so difficult for me to avoid overeating. From a boy I was taught to leave no scraps on my plate, and from this excellent general rule of conduct I now suffer in my old age!...—Yours very sincerely,

ALFRED R. WALLACE.

### TO DR. LITTLEDALE

*Old Orchard, Broadstone, Dorset. January 11, 1913.*

Dear Dr. Littledale,—Many thanks for your kind congratulations and good wishes.[48] I am glad to say I feel still able to jog on a few years longer in this *very good* world—for those who can make the best of it.

I am now suffering most from "eczema," which has settled in my legs, so that I cannot stand or walk for any length of time. Perhaps that is an outlet for something worse, as I still enjoy my meals, and usually feel as well as ever, though I have to be very careful as to *what* I eat.—With best wishes for your prosperity, yours very truly,

ALFRED R. WALLACE.

### TO DR. NORRIS

*Old Orchard, Broadstone, Dorset. October 4, 1913.*

My dear Dr. Norris,—Except for a continuous weakness I seem improving a little in general health, and the chronic rheumatic pain in my right shoulder has almost passed away in the last month (after about three years), and I can impute it to nothing but about a quarter of a pint a day of Bulmer's Cider! A most agreeable medicine!

The irritability of the skin, however, continues, though the inflammation of the legs has somewhat diminished....

My increasing weakness is now my most serious trouble, as it prevents me really from doing any more work, and causes a large want of balance, and liability to fall down. Even moving about the room after books, etc., dressing and undressing, make me want to lie down and rest....

With kind remembrances to your daughter, believe me yours very sincerely,

In disposition Dr. Wallace was cheerful, and very optimistic, and remarkably even-tempered. If irritated he quickly recovered, and soon forgot all about the annoyance, but he was always strongly indignant at any injustice to the weak or helpless. When worried by business difficulties or losses he very soon recovered his optimism, and seemed quite confident that all would come right (as indeed it generally did), and latterly he became convinced that all his past troubles were really blessings in disguise, without which as a stimulant he would have done no useful work.

His life was a happy one, and even the discomforts caused by his ailments, which were at times very acute for days together, never prevented him from enjoying the contemplation of his flowers, nor disturbed the serenity of his temper, nor caused him to complain.

Although rather delicate all his life, he rarely stayed in bed; in fact, only once in our memory, during an illness at Parkstone, did he do so, and then only for one day.

On Saturday, November 1st (1913), he walked round the garden, and on the following day seemed very bright, and enjoyed his dinner and supper, but about nine o'clock he felt faint and shivered violently. We called in Dr. Norman, who came in about an hour, and we heard them having a long talk and even laughing, in the study. As the doctor left he said, "Wonderful man! he knows so much. I can do nothing for him."

The next day he did not get up at the usual time, but we felt no anxiety until noon, when he still showed no inclination to rise. He appeared to be dozing, and said he wanted nothing. From that time he gradually sank into semi-consciousness, and at half-past nine in the morning of Friday, November 7th, quietly passed on to that other life in which he was such a firm believer.

# PART V

## SOCIAL AND POLITICAL VIEWS

> "When a country is well governed, poverty and a mean condition are things to be ashamed of. When a country is ill governed, riches and honour are things to be ashamed of."—CONFUCIUS.

In the above sentences, written long before the dawn of Christian civilisation, we have an apt summary of the social and political views of Alfred Russel Wallace.

As we have stated in a previous chapter, it was during his short stay in London as a boy, when he was led to study the writings and methods of Robert Owen, of New Lanark, that his mind first opened to the consideration of the inequalities of our social life.

During the six years which he spent in land-surveying he obtained a more practical knowledge of the laws pertaining to public and private property as they affected the lives and habits of both squire and peasant.

The village inn, or public-house, was then the only place where men could meet to discuss topics of mutual interest, and it was there that young Wallace and his brother spent some of their own leisure hours listening to and conversing with the village rustics. The conversation was not ordinarily of an educational character, but occasionally experienced farmers would discuss agricultural and land problems which were beginning to interest Wallace.

In reading his books and essays written more than seventy years later, we are struck with the exceptional opportunities which he had of comparing social conditions, and commercial and individual prosperity during that long period, and of witnessing the introduction of many inventions. He used to enjoy recalling many of the discussions between intelligent mechanics which he heard of in his early days regarding the introduction of the steam-engine. One and another declared that the grip of the engine on the rails would not be sufficient to draw heavy trucks or carriages; that the wheels, in fact, would whiz round instead of going on, and that it would be necessary to sprinkle sand in front of the wheels, or make the tyres rough like files. About this time, too, there arose a keen debate upon the relative merits of the new railroads and the old canals. Many thought that the former could never compete with the latter in carrying heavy goods; but facts soon proved otherwise, for in one district alone the traffic of the canal, within two years of the coming of the railway, decreased by 1,000,000 tons.

It was during these years, and when he and his brother were making a survey for the enclosure of some common lands near Llandrindod Wells, that Wallace finally became aware of the injustice towards the labouring classes of the General Enclosure Act.

In this particular locality the land to be enclosed consisted of a large extent of moor, and mountain which, with other common rights, had for many years enabled the occupants of the scattered cottages around to keep a horse, cow, or a few sheep, and thus make a fairly comfortable living. Under the Act, the whole of this open land was divided among the adjacent landowners of the parish or manor, in proportion to the size or value of their estates. Thus, to those who actually possessed much, much was given; whilst to those who only nominally owned a little land, even that was taken away in return for a small compensation which was by no means as valuable to them as the right to graze their cattle. In spite of the statement set forth in the General Enclosure Act—"Whereas it is expedient to facilitate the enclosure and improvement of common and other lands now subject to the rights of property which obstruct cultivation and the productive employment of labour," Wallace ascertained many years later that no single part of the land so enclosed had been cultivated by those to whom it was given, though certain portions had been let or sold at fabulous prices for building purposes, to accommodate summer visitors to the neighbourhood. Thus the unfortunate people who had formerly enjoyed home, health, and comparative prosperity in the cottages scattered over this common land had been obliged to migrate to the large towns, seeking for fresh employment and means of subsistence, or had become "law-created paupers"; whilst to crown all, the piece of common originally "reserved" for the benefit of the inhabitants had been turned into golf-links!

Again and again Wallace drew attention to the fundamental duties of landownership, maintaining that the public, as a whole, had become so blinded by custom that no effectual social reform would ever be established unless some strenuous and unremitting effort was made to recover the land by law from those who had made the land laws and who had niched the common heritage of humanity for their own private aggrandisement.

With regard to the actual value of land, Wallace pointed out that the last valuation was made in the year 1692, and therefore, with the increase of value through minerals and other products since then, the arrears of land tax due up to 1905 would amount to more than the value of all the agricultural land of our country at the present time; therefore existing landlords, in clamouring for their alleged rights of property, might find out that those "rights" no longer exist.

Yet another point on which he insisted was the right of way through fields or woodlands, and especially beside the sea. With the advent of the motor-car and other swift means of locomotion, the public roads are no longer safe and pleasurable for pedestrians; besides the iniquitous fact that hundreds are kept from enjoying the beauties of nature by the utterly selfish and useless reservations of such by-paths by the landowner.

"This all-embracing system of land-robbery," again he writes, "for which nothing is too great or too small; which has absorbed meadow and forest, moor and mountain, which has appropriated most of our rivers and lakes and the fish that live in them; making the agriculturist pay for his seaweed manure and the fisherman for his bait of shell-fish; which has desolated whole counties to replace men by sheep or cattle, and has destroyed fields and cottages to make a wilderness for deer and grouse; which has stolen the commons and filched the roadside wastes; which has driven the labouring poor into the cities, and thus been the chief cause of the misery, disease, and early death of thousands ... it is the advocates of this inhuman system who, when a partial restitution of their unholy gains is proposed, are the loudest in their cries of 'robbery'!

"But all the robbery, all the spoliation, all the legal and illegal filching, has been on *their* side.... They made the laws to legalise their actions, and, some day, we, the people, will make laws which will not only legalise but justify our process of restitution. It will justify it, because, unlike their laws, which always took from the poor to give to the rich—to the very class which made the laws—ours will only take from the superfluity of the rich, *not* to give to the poor or to any individuals, but to so administer as to enable every man to live by honest work, to restore to the whole people their birthright in their native soil, and to relieve all alike from a heavy burden of unnecessary and unjust taxation. *This* will be the true statesmanship of the future, and it will be justified alike by equity, by ethics, and by religion."

These, then, are the facts and reasons upon which Dr. Wallace based his strenuous advocacy of Land Nationalisation.[42] It was only by slow degrees that he arrived at some of the conclusions propounded in his later years, but once having grasped their full importance to the social and moral well-being of the community, he held them to the last.

The first book which tended to fasten his attention upon these matters was "Social Statics," by Herbert Spencer, but in 1870 the publication of his "Malay Archipelago" brought him into personal contact with John Stuart Mill, through whose invitation he became a member of the General Committee of the Land Tenure Reform Association. On the formation of the Land Nationalisation Society in 1880 he retired from the Association, and devoted himself to the larger issues which the new Society embraced.

Soon after the latter Society was started, Henry George, the American author of "Progress and Poverty," came to England, and Wallace had many opportunities of hearing him speak in public and of discussing matters of common interest in private. In spite of the ridicule poured upon Henry George's book by many eminent social reformers, Wallace consistently upheld its general principles.

His second work on these various subjects was a small book entitled "Bad Times," issued in 1885, in which he went deeply into the root causes of the depression in trade which had lasted since 1874. The facts there given were enlarged upon and continually brought up to date in his later writings. Articles which had appeared in various magazines were gathered together and included, with those on other subjects, in "Studies, Scientific and Social." His last three books, which include his ideas on social diseases and the best method of preventing them, were "The Wonderful Century," "Social Environment and Moral Progress," and "The Revolt of Democracy"; the two last being issued, as we have seen, in 1913, the year of his death.

In "Social Environment and Moral Progress" the conclusion of his vehement survey of our moral and social conditions was startling: "*It is not too much to say that our whole system of Society is rotten from top to bottom, and that the social environment as a whole in relation to our possibilities and our claims is the worst that the world has ever seen.*"

That terrible indictment was doubly underscored in his MS.

What, in his mature judgment, were the causes and remedies? He set them out in this order:

1. The evils are due, broadly and generally, to our living under a system of universal competition for the means of existence, the remedy for which is equally universal co-operation.

2. It may also be defined as a system of economic antagonism, as of enemies, the remedy being a system of economic brotherhood, as of a great family, or of friends.

3. Our system is also one of monopoly by a few of all the means of existence—the land, without access to which no life is possible; and capital, or the results of stored-up labour, which is now in the possession of a limited number of capitalists, and therefore is also a monopoly. The remedy is freedom of access to land and capital for all.

4. Also, it may be defined as social injustice, inasmuch as the few in each generation are allowed to inherit the stored-up wealth of all preceding

generations, while the many inherit nothing. The remedy is to adopt the principle of equality of opportunity for all, or of universal *inheritance by the State in trust for the whole community.*

"We have," he finally concluded, "ourselves created an immoral or unmoral social environment. To undo its inevitable results we must reverse our course. We must see that *all* our economic legislation, *all* our social reforms, are in the very opposite direction to those hitherto adopted, and that they tend in the direction of one or other of the four fundamental remedies I have suggested. In this way only can we hope to change our existing immoral environment into a moral one, and *initiate a new era of Moral Progress.*" The "Revolt of Democracy"[50] was addressed directly to the Labour Party. And once again he drew a vivid picture of how, during the whole of the nineteenth century, there was a continuous advance in the application of scientific discovery to the arts, especially to the invention and application of labour-saving machinery; and how our wealth had increased to an equally marvellous extent.

He pointed out that various estimates which had been made of the increase in our wealth-producing capacity showed that, roughly speaking, the use of mechanical power had increased it more than a hundredfold during the century; yet the result had been to create a limited upper class, living in unexampled luxury, while about one-fourth of the whole population existed in a state of fluctuating penury, often sinking below the margin of poverty. Many thousands were annually drawn into this gulf of destitution, and died from direct starvation and premature exhaustion or from diseases produced by unhealthy employment.

During this long period, however, although wealth and want had alike increased side by side, public opinion had not been sufficiently educated to permit of any effectual remedy being applied. The workers themselves had failed to visualise its fundamental causes, land monopoly and the competitive system of industry giving rise to an ever-increasing private capitalism which, to a very large extent, had controlled the Legislature. All through the last century this rapid accumulation of wealth due to extensive manufacturing industries led to a still greater increase of middlemen engaged in the distribution of the products, from the wealthy merchant to the various grades of tradesmen and small shop-keepers who supplied the daily wants of the community.

To those who lived in the midst of this vast industrial system, or were a part of it, it seemed natural and inevitable that there should be rich and poor; and this belief was enforced on the one hand by the clergy, and on the other by political economists, so that religion and science agreed in upholding the competitive and capitalistic system of society as the only rational and possible one. Hence it came to be believed that the true sphere of governmental action did not include the abolition of poverty. It was even declared that poverty was due to economic causes over which governments had no power; that wages were kept down by the "iron law" of supply and demand; and that any attempt to find a remedy by Acts of Parliament only aggravated the disease. During the Premiership of Sir Henry Campbell-Bannerman this attitude was, for the first time, changed. On numerous occasions Sir Henry declared that he held it to be the duty of a government to deal with problems of unemployment and poverty.

In 1908 three great strikes, coming in rapid succession—those of the Railway and other Transport Unions, the Miners, and the London Dock Labourers—brought home to the middle and upper classes, and to the Government, how completely all are dependent on the "working classes." This and similar experiences showed us that when the organisation of the trade unions was more complete, and the accumulated funds of several years were devoted to this purpose, the bulk of the inhabitants of London, and of other great cities, could be made to suffer a degree of famine comparable with that of Paris when besieged by the German army in 1870.

Wallace's watchword throughout these social agitations was "Equality of Opportunity for All," and the ideal method by which he hoped to achieve this end was a system of industrial colonisation in our own country whereby *all* would have a fair, if not an absolutely equal, share in the benefits arising from the production of their own labour, whether physical or mental.[51]

With regard to the education of the people, especially as a stepping-stone to moral and intellectual reform, Wallace believed in the training of individual natural talent, rather than the present system of general education thrust upon every boy or girl regardless of their varying mental capacities. He also urged that the building-up of the mind should be alternated with physical training in one or more useful trades, so that there might be, not only at the outset, but also in later life, a choice of occupation in order to avoid the excess of unemployment in any one direction.

In his opinion, one of the injurious results of our competitive system, having its roots, however, in the valuable "guilds" of a past epoch, was the almost universal restriction of our workers to only one kind of labour. The result was a dreadful monotony in almost all spheres of work, the extreme unhealthiness of many, and a much larger amount of unemployment than if each man or woman were regularly trained in two or more occupations. In addition to two of what are commonly called trades, every youth should be trained for one day a week or one week in a month, according to the demand for labour, in some of the various operations of farming or gardening. Not only would this improve the general health of the workers, but it would also add much to the interest and enjoyment of their lives.

"There is one point," he wrote, "in connection with this problem which I do not think has ever been much considered or discussed. It is the undoubted benefit to all the members of a society of *the greatest possible diversity of character*, as a means both towards the greatest enjoyment and interest of association, and to the highest ultimate development of the race. If we are to suppose that man might have been created or developed with none of those extremes of character which now often result in what we call wickedness, vice, or crime, there would certainly have been a greater monotony in human nature, which would, perhaps, have led to less beneficial results than the variety which actually exists may lead to. We are more and more getting to see that very much, perhaps all, the vice, crime, and misery that exists in the world is the result, not of the wickedness of individuals, but of the entire absence of sympathetic training from infancy onwards. So far as I have heard, the only example of the effects of such a training on a large scale was that initiated by Robert Owen at New Lanark, which, with most unpromising materials, produced such marvellous results on the character and conduct of the children as to seem almost incredible to the numerous persons who came to see and often critically to examine them. There must have been all kinds of characters in his schools, yet *none* were found to be incorrigible, *none* beyond control, *none* who did not respond to the love and sympathetic instruction of their teachers. It is therefore quite possible that *all* the evil in the world is directly due to man, not to God, and that when we once realise this to its full extent we shall be able, not only to eliminate almost completely what we now term evil, but shall then clearly perceive that all those propensities and passions that under bad conditions of society inevitably led to it, will under good conditions add to the variety and the capacities of human nature, the enjoyment of life by all, and at the same time greatly increase the possibilities of development of the whole race. I myself feel confident that this is really the case, and that such considerations, when followed out to their ultimate issues, afford a complete solution of the great problem of the ages—the origin of evil."[52]

Closely allied with the welfare of the child is another "reform" with which Wallace's name will long be associated. That is his strong denunciation of Vaccination. For seven years he laboured to show medical and scientific men that statistics proved beyond doubt the futility of this measure to prevent disease. A few were converted, but public opinion is hard to move.

In his ideal of the future, Dr. Wallace gave a large and honoured sphere to women. He considered that it was in the highest degree presumptuous and irrational to attempt to deal by compulsory enactments with the most vital and most sacred of all human relationships, regardless of the fact that our present phase of social development is not only extremely imperfect, but, as already shown, vicious and rotten to the core. How could it be possible to determine by legislation those relations of the sexes which shall be best alike for individuals and for the race in a society in which a large proportion of our women are forced to work long hours daily for the barest subsistence, with an almost total absence of the rational pleasures of life, for the want of which thousands are driven into uncongenial marriages in order to secure some amount of personal independence or physical well-being. He believed that when men and women are, for the first time in the course of civilisation, equally free to follow their best impulses; when idleness and vicious and hurtful luxury on the one hand, and oppressive labour and the dread of starvation on the other, are alike unknown; when *all* receive the best and broadest education that the state of civilisation and knowledge will admit; when the standard of public opinion is set by the wisest and the best among us, and that standard is systematically inculcated in the young—then we shall find that a system of truly "Natural Selection" (a term that Wallace preferred to "Eugenics," which he utterly disliked) will come spontaneously into action which will tend steadily to eliminate the lower, the less developed, or in any way defective types of men, and will thus continuously raise the physical, moral, and intellectual standard of the race.

He further held that "although many women now remain unmarried from necessity rather than from choice, there are always considerable numbers who feel no strong impulse to marriage, and accept husbands to secure subsistence and a home of their own rather than from personal affection or sexual emotion. In a state of society in which all women were economically independent, where all were fully occupied with public duties and social or intellectual pleasures, and had nothing to gain by marriage as regards material well-being or social position, it is highly probable that the numbers of unmarried from choice would increase. It would probably come to be considered a degradation for any woman to marry a man whom she could not love and esteem, and this reason would tend at least to delay marriage till a worthy and sympathetic partner was encountered."

But this choice, he considered, would be further strengthened by the fact that, with the ever-increasing approach to equality of opportunity for every child born in our country, that terrible excess of male deaths, in boyhood and early manhood especially, due to various preventable causes, would disappear, and change the present majority of women to a majority of men. This would lead to a greater rivalry for wives, and give to women the power of rejecting all the lower types of character among their suitors.

"It will be their special duty so to mould public opinion, through home training and social influence, as to render the women of the future the regenerators of the entire human race." He fully hoped and believed that they would prove equal to the high and responsible position which, in accordance with natural laws, they will be called upon to fulfil.

Mr. D.A. Wilson, who visited him in 1912, writes:

He surprised me by saying he was a Socialist—one does not expect a man like him to label himself in any way. It appeared to be unconscious modesty, like a school-boy's, which made him willing to be labelled; but no label could describe him, and his mental sweep was unlimited. Although in his ninetieth year, he seemed to be in his prime. There was no sign of age but physical weakness, and you had to make an effort at times to remember even that. His eye kindled as he spoke, and more than once he walked about and chuckled, like a schoolboy pleased.

An earnest expression like Carlyle's came over his countenance as he reprobated the selfish, wild-cat competition which made life harder and more horrible to-day for a well-doing poor man in England than among the Malays or Burmese before they had any modern inventions. Co-operation was the upward road for humanity. Men grew out of beasthood by it, and by it civilisation began. Forgetting it, men retrograded, subsiding swiftly, so that there were many individuals among us to-day who were in body, mind, and character below the level of our barbarian ancestors or contemporary "savages," to say nothing of civilised Burmese or Malays. What he meant by Socialism can be seen from his books. Nothing in them surprised me after our talk. His appreciation of Confucius, when I quoted some things of the Chinese sage's which confirmed what he was saying, was emphatic, and that and many other things showed that Socialism to him implied the upward evolution of humanity. It was because of the degradation of men involved that he objected to letting individuals grab the public property—earth, air and water. Monopolies, he thought, should at once revert to the public, and we had an argument which showed that he had no objection to even artificial monopolies if they were public property. He defended the old Dutch Government monopolies of spices, and declared them better than to-day's free trade, when cultivation is exploited by men who always tended to be mere money-grabbers, selfish savages let loose. In answer I mentioned the abuses of officialdom, as seen by me from the inside in Burma, and he agreed that the mental and moral superiority of many kinds of Asiatics to the Europeans who want to boss them made detailed European administration an absurdity. We should leave these peoples to develop in their own way. Having conquered Burma and India, he proceeded, the English should take warning from history and restrict themselves to keeping the peace, and protecting the countries they had taken. They should give every province as much home rule as possible and as soon as possible, and study to avoid becoming parasites.—D.A.W.

We may fittingly conclude this brief summary of Wallace's social views and ideals by citing his own reply to the question: "Why am I a Socialist?" "I am a Socialist because I believe that the highest law for mankind is justice. I therefore take for my motto, 'Fiat Justitia, Ruat Coelum'; and my definition of Socialism is, 'The use, by everyone, of his faculties for the common good, and the voluntary organisation of labour for the equal benefit of all.' That is absolute social justice; that is ideal Socialism. It is, therefore, the guiding star for all true social reform."

He corresponded with Miss Buckley not only on scientific but also on public questions and social problems:

### TO MISS BUCKLEY

*Rosehill, Dorking. Sunday, [? December, 1878].*

Dear Miss Buckley,— ... How wonderfully the Russians have got on since you left! A very little more and the Turkish Government might be turned out of Europe—even now it might be with the greatest ease if our Government would join in giving them the last kick. Whatever power they retain in Europe will most certainly involve another war before twenty years are over.—Yours very faithfully,

ALFRED R. WALLACE.

### TO MISS BUCKLEY

*Waldron Edge, Croydon. May 2, 1879.*

Dear Miss Buckley,— ... My "Reciprocity" article seems to have produced a slight effect on the *Spectator*, though it did snub me at first, but it is perfectly sickening to read the stuff spoken and written, in Parliament and in all the newspapers, about the subject, all treating our present practice as something holy and immutable, whatever bad effects it may produce, and though it is not in any way "free trade" and would I believe have been given up both by Adam Smith and Cobden.—Yours very faithfully,

ALFRED R. WALLACE.

He was always ready, even eager, to discuss his social and land nationalisation principles with his scientific friends, with members of his own family, and indeed with anyone who would lend a willing ear.

### HERBERT SPENCER TO A.R. WALLACE

*38 Queen's Gardens, Bayswater, W. April 25, 1881.*

Dear Mr. Wallace,—As you may suppose, I fully sympathise with the general aims of your proposed Land Nationalisation Society; but for sundry reasons I hesitate to commit myself, at the present stage of the question, to a programme so definite as that which you send me. It seems to me that before formulating the idea in a specific shape it is needful to generate a body of public opinion on the general issue, and that it must be some time before there can be produced such recognition of the general principle involved as is needful before definite plans can be set forth to any purpose....—Truly yours,

HERBERT SPENCER.

### HERBERT SPENCER TO A.R. WALLACE

*38 Queen's Gardens, Bayswater, W. July 6, 1881.*

Dear Mr. Wallace,—I have already seen the work you name, "Progress and Poverty," having had a copy, or rather two copies, sent me. I gathered from what little I glanced at that I should fundamentally disagree with the writer, and have not read more.

I demur entirely to the supposition, which is implied in the book, that by any possible social arrangements whatever the distress which humanity has to suffer in the course of civilisation could have been prevented. The whole process, with all its horrors and tyrannies, and slaveries, and wars, and abominations of all kinds, has been an inevitable one accompanying the survival and spread of the strongest, and the consolidation of small tribes into large societies; and among other things the lapse of land into private ownership has been, like the lapse of individuals into slavery, at one period of the process altogether indispensable. I do not in the least believe that from the primitive system of communistic ownership to a high and finished system of State ownership, such as we may look for in the future, there could be any transition without passing through such stages as we have seen and which exist now. Argument aside, however, I should be disinclined to commit myself to any scheme of immediate action, which, as I have indicated to you, I believe at present premature. For myself I feel that I have to consider not only what I may do on special questions, but also how the action I take on special questions may affect my general influence; and I am disinclined to give more handles against me than are needful. Already, as you will see by the enclosed circular, I am doing in the way of positive action more than may be altogether prudent.—Sincerely yours,

HERBERT SPENCER.

### A.R. WALLACE TO MR. A.C. SWINTON

*Frith Hill, Godalming. December 23, 1885.*

My dear Swinton,— ... I have just received an invitation to go to lecture in Sydney on Sundays for three months, with an intimation that other lectures can be arranged for in Melbourne and New Zealand. It is tempting!... If I had the prospect of clearing £1,000 by a lecturing campaign I would go, though it would require a great effort.... I did not think it possible even to contemplate going so far again, but the chance of earning a lot of money which would enable me to clear off this house and leave something for my family must be seriously considered.—Yours very truly,

ALFRED R. WALLACE.

### TO Miss VIOLET WALLACE

*Parkstone, Dorset. May* 10, 1891.

My dear Violet,— ... I am quite in favour of a legal eight hours' day. Overtime need not be forbidden, but every man who works overtime should have a legal claim to double wages for the extra hours. That would make it cheaper for the master to employ two sets of men working each eight hours when they had long jobs requiring them, while for the necessities of finishing contracts, etc., they could well afford to pay double for the extra hours. "It would make everything dearer!" Of course it would! How else can you produce a more equal distribution of wealth than by making the rich and idle pay more and the workers receive more? "The workers would have to pay more, too, for everything they bought!" True again, but what they paid more would not equal their extra earnings, because a large portion of the extra pay to the men will be paid by the rich, and only the remainder paid by the men themselves. The eight hours' day and double pay for overtime would not only employ thousands now out of work, but would actually raise wages per hour and per day. This is clear, because wages are kept down wholly by the surplus supply of labour in every trade. The moment the surplus is used up, or nearly so, by more men being required on account of shorter hours, competition among the men becomes less; among the employers, for men, more: hence necessarily higher wages all round. As to the bogey of foreign competition, it is a bogey only. All the political economists agree that if wages are raised in all trades, it will not in the least affect our power to export goods as profitably as now. Look and see! And, secondly, the eight hours' movement is an international one, and will affect all alike in the end.

There are some arguments for you! Poor unreasoning infant!!...

## REV. AUGUSTUS JESSOPP TO A.R. WALLACE

*Scarning Rectory, East Dereham. August 25, 1893.*

My dear Mr. Wallace,—I have put off writing to thank you for your kind letter, and the book and pamphlets you were good enough to send me, because I hoped in acknowledgment to say I had read your little volumes, as I intend to. The fates have been against me, and I will delay no longer thanking you for sending them to me.

I do not believe in your theory of land nationalisation one bit! But I like to see all that such a man as you has to say on his side.

In return I send you my view of the matter, which is just as likely to convert you as your book is to convert me.

I love a man with a theory, for I learn most from such a man, and when I have thought a thing out in my own mind and forgotten the arguments while I have arrived at a firm conviction as to the conclusion, it is refreshing to be reminded of points and facts that have slipped away from me!

It was a great pleasure and privilege to make your acquaintance the other day, and I hope we may meet again some day.—Very truly yours,

AUGUSTUS JESSOPP.

## REV. H. PRICE HUGHES TO A.R. WALLACE

*8 Taviton Street, Gordon Square, W.C. September 14, 1898.*

Dear Dr. Wallace,—I am always very glad when I hear from you. So far as your intensely interesting volume has compelled some very prejudiced people to read your attack on modern delusions, it is a great gain, especially to themselves. I have read your tract on "Justice, not Charity," with great pleasure and approval. The moment Mr. Benjamin Kidd invented the striking term of "equality of opportunity" I adopted it, and have often preached it in the pulpit and on the platform, just as you preach it in the tract before me. I fully agree that justice, not charity, is the fundamental principle of social reform. There is something very contemptible in the spiteful way in which many newspapers and magistrates are trying to aggravate the difficulties of conscientious men who avail themselves of the conscience clause in the new Vaccination Act. There is very much to be done yet before social justice is realised, but the astonishing manifesto of the Czar of Russia, which I have no doubt is a perfectly sincere one, is a revelation of the extent to which social truth is leavening European society. Since I last wrote to you I have been elected President of the Wesleyan Methodist Conference, which will give me a great deal of special work and special opportunities also, I am thankful to say, of propagating Social Christianity, which in fact, and to a great extent in form, is what you yourself are doing.—Yours very sincerely,

H. PRICE HUGHES.

## TO ALFRED RUSSELL

*Parkstone, Dorset. May 11, 1900.*

Dear Sir,—I am not a vegetarian, but I believe in it as certain to be adopted in the future, and as essential to a higher social and moral state of society. My reasons are:

(1) That far less land is needed to supply vegetable than to supply animal food.

(2) That the business of a butcher is, and would be, repulsive to all refined natures.

(3) That with proper arrangements for variety and good cookery, vegetable food is better for health of body and mind.—Yours very truly,

ALFRED R. WALLACE.

## TO MR. JOHN (LORD) MORLEY

*Parkstone, Dorset, October 20, 1900.*

Dear Sir,—I look upon you as the one politician left to us, who, by his ability and integrity, his eloquence and love of truth, his high standing as a thinker and writer, and his openness of mind, is able to become the leader of the English people in their struggle for freedom against the monopolists of land, capital, and political power. I therefore take the liberty of sending you herewith a book of mine containing a number of miscellaneous essays, a few of which, I venture to think, are worthy of your serious attention.

Some time since you intimated in one of your speeches that, if the choice for this country were between Imperialism and Socialism, you were inclined to consider the latter the less evil of the two. You added, I think, your conviction that the dangers of Socialism to human character were what most influenced you against it. I trust that my impression of what you said is substantially correct. Now I myself believe, after a study of the subject extending over twenty years, that this danger is non-existent, and certainly does not in any way apply to the fundamental principles of Socialism, which is, simply, *the voluntary organisation of labour for the good of all*....—With great esteem, I am yours very faithfully,

ALFRED R. WALLACE.

## MR. JOHN (LORD) MORLEY TO A.R. WALLACE

*57 Elm Park Gardens, S.W. October 31, 1900.*

My dear Sir,—For some reason, though your letter is dated the 20th, it has only reached me, along with the two volumes, to-day. I feel myself greatly indebted to you for both. In older days I often mused upon a passage of yours in the "Malay Archipelago" contrasting the condition of certain types of savage life with that of life in a modern industrial city. And I shall gladly turn again to the subject in these pages, new to me, where you come to close quarters with the problem.

But my time and my mind are at present neither of them free for the effective consideration of this mighty case. Nor can I promise myself the requisite leisure for at least several months to come. What I can do is to set your arguments a-simmering in my brain, and perhaps when the time of liberation arrives I may be in a state to make something of it. I don't suppose that I shall be a convert, but I always remember J.S. Mill's observation, after recapitulating the evils to be apprehended from Socialism, that he would face them in spite of all, if the only alternative to Socialism were our present state.—With sincere thanks and regard, believe me yours faithfully

JOHN MORLEY.

## TO MR. C.G. STUART-MENTEITH

*Parkstone, Dorset. June 6, 1901.*

Dear Sir,—I have no time to discuss your letter[53] at any length. You seem to assume that we can say definitely who are the "fit" and who the "unfit."

I deny this, except in the most extreme cases.

I believe that, even now, the race is mostly recruited by the *more fit*—that is the upper working classes and the lower middle classes.

Both the very rich and the very poor are probably—as classes—below these. The former increase less rapidly through immorality and late marriage; the latter through excessive infant mortality. If that is the case, no legislative interference is needed, and would probably do harm.

I see nothing in your letter which is really opposed to my contention—that under rational social conditions the healthy instincts of men and women will solve the population problem far better than any tinkering interference either by law or by any other means.

And in the meantime the condition of things is not so bad as you suppose.—Yours very truly,

ALFRED R. WALLACE.

## TO MR. SYDNEY COCKERELL

*Broadstone, Wimborne. January 15, 1906.*

Dear Mr. Cockerell,—I have now finished reading Kropotkin's Life with very great interest, especially for the light it throws on the present condition of Russia. It also brings out clearly some very fine aspects of the Russian character, and the horrible despotism to which they are still subject, equivalent to that of the days of the Bastille and the system of *Lettres de cachet* before the great Revolution in France. It seems to me probable that under happier conditions—perhaps in the not distant future—Russia may become the most advanced instead of the most backward in civilisation—a real leader among nations, not in war and conquest but in social reform.—Yours faithfully,

A.R. WALLACE.

## TO MR. J. HYDER (Of THE LAND NATIONALISATION SOCIETY)

*Broadstone, Wimborne. May 13, 1907.*

Dear Mr. Hyder,—Although it is not safe to hallo before one is out of the wood, I think I may congratulate the Society upon the prospect it now has of obtaining the first-fruits of its persistent efforts, for a quarter of a century, to form an enlightened public opinion in favour of our views. If the Government adequately fulfils its promises, we shall have, in the Bill for a fair valuation of land apart from improvements, as a basis of taxation and for purchase, and that giving local authorities full powers to acquire land so valued, the first real and definite steps towards complete nationalisation....

ALFRED R. WALLACE.

## TO MR. A. WILTSHIRE[54]

*Broadstone, Wimborne. October 10, 1907.*

Dear Sir,—I told Mr. Button that I do not approve of the resolution you are going to move.[55]

The workers of England have themselves returned a large majority of ordinary Liberals, including hundreds of capitalists, landowners, manufacturers, and lawyers, with only a sprinkling of Radicals and Socialists. The Government—your own elected Government—is doing more for the workers than any Liberal Government ever did before, yet you are going to pass what is practically a vote of censure on it for not being a Radical, Labour, and Socialist Government!

If this Government attempted to do what you and I think ought to be done, it would lose half its followers and be turned out, ignominiously, giving the Tories another chance. That is foolish as well as unfair.—Yours truly,

ALFRED R. WALLACE.

## TO LORD AVEBURY

*Broadstone, Wimborne. June 23, 1908.*

Dear Lord Avebury,— ... Allow me to wish every success to your Bill for preserving beautiful birds from destruction. To stop the import is the only way—short of the still more drastic method of heavily fining everyone who wears feathers in public, with imprisonment for a second offence. But we are not yet ripe for that.—Yours very truly,

## TO MR. E. SMEDLEY

*Old Orchard, Broadstone, Dorset. December 25, 1910.*

Dear Mr. Smedley,—Thanks for your long and interesting letter.... Man is, and has been, horribly cruel, and it is indeed difficult to explain why. Yet that there is an explanation, and that it does lead to good in the end, I believe. Praying is evidently useless, and should be, as it is almost always selfish—for *our* benefit, or our *families*, or our *nation*.—Yours very truly,

ALFRED R. WALLACE.

## TO MR. W.G. WALLACE

*Old Orchard, Broadstone, Wimborne. August 20, 1911.*

My dear Will,— ... The railway strike surpasses the Parliament Bill in excitement. On receipt of Friday's paper, I sat down and composed and sent off to Lloyd George a short but big letter, on large foolscap paper, urging him and Asquith, as the two strong men of the Government, to take over at once the management of the railways of the entire country, by Royal Proclamation—on the ground of mismanagement for seventy years, and having brought the country to the verge of starvation and civil war; to grant an amnesty to all strikers (except for acts of violence), also grant all the men's demands for one year, and devote that time to a deliberate and impartial inquiry and a complete scheme of reorganisation of the railways in the interest, first of the public, then of the men of all grades, lastly of the share and bond owners, who will become guaranteed public creditors.... It has been admitted and proved again and again, that the men are badly treated, that their grievances are real—their very unanimity and standing by each other proves it. Their demands are most moderate; and the cost in extra wages will be saved over and over in safety, regularity, economy of working, and public convenience. I have not had even an acknowledgment of receipt yet, but hope to in a day or two....

## MR. H.M. HYNDMAN TO A.R. WALLACE

*9 Queen Anne's Gate, Westminster, S.W. March 14, 1912.*

Dear Sir,—Everyone who knows anything of the record of modern science in this country recognises how very much we all owe to you. It was, therefore, specially gratifying to me that you should be so kind as to write such a very encouraging letter on the occasion of my seventieth birthday. I owe you sincere thanks for what you said, though I may honestly feel that you overpraised what I have done. It has been an uphill fight, but I am lucky in being allowed to see through the smoke and dust of battle a vision of the promised land. The transformation from capitalism to socialism is going on slowly under our eyes.

Again thanking you and wishing you every good wish, believe me yours sincerely,

H.M. HYNDMAN.

## TO MR. M.J. MURPHY

*Old Orchard, Broadstone, Dorset. August 19, 1913.*

Dear Sir,—I not only think but firmly believe that Lloyd George is working for the good of the people, in all ways open to him. The wonder is that he can persuade Asquith and the Cabinet to let him go as far as he does. No doubt he is obliged to do things he does not think the best absolutely, but the best that are practicable. He does not profess to be a Socialist, and he is not infallible, but he does the best he can, under the conditions in which he finds himself. Socialists who condemn him for not doing more are most unfair. They must know, if they think, that if he tried to do much more towards Socialism he would break up the Government and let in the Tories.—Yours truly,

A.R. WALLACE.

## TO MR. A. WILTSHIRE

*Old Orchard, Broadstone, Dorset. September 14, 1913.*

Dear Sir,—I wish you every success in your work for the amelioration of the condition of the workers, through whose exertions it may be truly said we all live and move and have our being.

Your motto is excellent. Above all things stick together.

Equally important is it to declare as a fixed principle that wages are to be and must be continuously raised, never lowered. You have too much arrears to make up—too many forces against you, to admit of their being ever lowered. Let future generations decide when that is necessary—if ever.

This is a principle worth enforcing by a general strike. Nothing less will be effective—nothing less should be accepted; and you must let the Government know it, and insist that they adopt it.

The rise must always be towards uniformity of payment for all useful and productive work.—Yours sincerely,

ALFRED R. WALLACE.

# PART VI

## Some Further Problems

### I.—Astronomy

Of the varied subjects upon which Wallace wrote, none, perhaps, came with greater freshness to the general reader than his books written when he was nearly eighty upon the ancient science of astronomy.

Perhaps he would have said that the "directive Mind and Purpose" kept these subjects back until the closing years of his life in order that he might bring to bear upon them his wider knowledge of nature, enlightened by that spiritual perception which led him to link the heavens and the earth in one common bond of evolution, culminating in the development of moral and spiritual intelligences.

"Man's Place in the Universe" (1903) was in effect a prelude to "The World of Life" (1910). Wallace saw afterwards that one grew out of the other, as we find him frequently saying with regard to his other books and essays.

As with Spiritualism, so with Astronomy, the seed-interest practically lay dormant in his mind for many years; with this difference, however, that temperament and training caused a speedy unfolding of his mind when once a scientific subject gripped him, whereas with Spiritualism he felt the need of moving slowly and cautiously before fully accepting the phenomena as verifiable facts.

It was during the later period of his land-surveying, when he was somewhere between the ages of 18 and 20, that he became distinctly interested in the stars. Being left much alone at this period, he began to vary his pursuits by studying a book on Nautical Astronomy, and constructing a rude telescope.[56] This primitive appliance increased his interest in other astronomical instruments, and especially in the grand onward march of astronomical discovery, which he looked upon as one of the wonders of the nineteenth century.

It was the inclusion of astronomy in lectures he delivered at Davos which led him to extend his original brief notes into the four chapters which form an important part of his "Wonderful Century." He freely confessed that in order to write these chapters he was obliged to read widely, and to make much use of friends to whom astronomy was a more familiar study. And it was whilst he was engaged upon these chapters that his attention became riveted upon the unique position of our planet in relation to the solar system.

He had noticed that certain definite conditions appeared to be absolutely essential to the origin and development of the higher types of terrestrial life, and that most of these must have been certainly dependent on a very delicate balance of the forces concerned in the evolution of our planet. Our position in the solar system appeared to him to be peculiar and unique because, he thought, we may be almost sure that these conditions do not coexist on any other planet, and that we have no good reason to believe that other planets could have maintained over a period of millions of years the complex and equable conditions absolutely necessary to the existence of the higher forms of terrestrial life. Therefore it appeared to him to be proved that our earth does really stand alone in the solar system by reason of its special adaptation for the development of human life.

Granting this, however, the question might still be asked, Why should not any one of the suns in other parts of space possess planets as well adapted as our own to develop the higher forms of organic life? These questions cannot be answered definitely; but there are reasons, he considered, why the central position which we occupy may alone be suitable. It is almost certain that electricity and other mysterious radiant forces (of which we have so recently discovered the existence) have played an important part in the origin and development of organised life, and it does not appear to be extravagant to assume that the extraordinary way in which these cosmic forces have remained hidden from us may be due to that central position which we are found to occupy in the whole universe of matter discoverable by us. Indeed, it may well be that these wonderful forces of the ether are more irregular—and perhaps more violent—in their effect upon matter in what may be termed the outer chambers of that universe, and that they are only so nicely balanced, so uniform in their action, and so concealed from us, as to be fit to aid in the development of organic life in that central portion of the stellar system which our globe occupies. Should these views as to the unique central position of our earth be supported by the results of further research, it will certainly rank as the most extraordinary and perhaps the most important of the many discoveries of the past century.

While still working on this section of his "Wonderful Century," he was asked to write a scientific article, upon any subject of his own choice, for the *New York Independent*. And as the idea of the unique position of the earth to be the abode of human life was fresh in his mind, he thought it would prove interesting to the general public. However, before his article appeared simultaneously in the American papers and in the *Fortnightly Review*, a friend who read it was so impressed with its originality and treatment that he persuaded Wallace to enlarge it into book form; and it appeared in the autumn of 1903 as "Man's Place in the Universe."

This fascinating treatise upon the position occupied by the earth, and man, in the universe, had the same effect as some of his former writings, of drawing forth unstinted commendation from many religious and secular papers; whilst the severely scientific and materialistic reviewers doubted how far his imagination had superseded unbiased reason.

On one point, however, most outsiders were in agreement—that he had invested an ancient subject with freshest interest through approaching it by an entirely new way. The plan followed was that of bringing together all the positive conclusions of the astronomer, the geologist, the physicist, and the biologist, and by weighing these carefully in the balance he arrived at what appeared to him to be the only reasonable conclusion. He therefore set out to solve the problem whether or not the logical inferences to be drawn from the various results of modern science lent support to the view that our earth is the only inhabited planet, not only in our own solar system, but in the whole stellar universe. In the course of his close and careful exposition he takes the reader through the whole trend of modern scientific research, concluding with a summing-up of his deductions in the following six propositions, in the first three of which he sets out the conclusions reached by modern astronomers:

(1) That the stellar universe forms one connected whole; and, though of enormous extent, is yet finite, and its extent determinable.

(2) That the solar system is situated in the plane of the Milky Way, and not far removed from the centre of that plane. The earth is, therefore, nearly in the centre of the stellar universe.

(3) That this universe consists throughout of the same kinds of matter, and is subjected to the same physical and chemical laws.

The conclusions which I claim to have shown to have enormous probabilities in their favour are:

(4) That no other planet in the solar system than our earth is inhabited or habitable.

(5) That the probabilities are almost as great against any other sun possessing inhabited planets.

(6) That the nearly central position of our sun is probably a permanent one, and has been specially favourable, perhaps absolutely essential, to life-development on the earth.

Wallace never maintained that this earth alone in the whole universe is the abode of life. What he maintained was, first, that our solar system appears to be in or near the centre of the visible universe, and, secondly, that all the available evidence supports the idea of the extreme unlikelihood of there being on any star or planet revealed by the telescope any intelligent life either identical with or analogous to man. To suppose that this one particular type of universe extends over all space was, he considered, to have a low idea of the Creator and His power. Such a scheme would mean monotony instead of infinite variety, the keynote of things as they are known to us. There might be a million universes, but all different.

To his mind there was no difficulty in believing in the existence of consciousness apart from material organism; though he could not readily conceive of pure mind, or pure spirit, apart from some kind of substantial envelope or substratum. Many of the views suggested in "Man's Place in the Universe" as to man's spiritual progress hereafter, the reason or ultimate purpose for which he was brought into existence, were enlarged upon, later, in "The World of Life." As early, however, as 1903, Wallace did not hesitate to express his own firm conviction that Science and Spiritualism were in many ways closely akin.

He believed that the near future would show the strong tendency of scientists to become more religious or spiritual. The process, he thought, would be slow, as the general attitude has never been more materialistic than now. A few have been bold enough to assert their belief in some outside power, but the leading scientific men are, as a rule, dead against them. "They seem," he once remarked, "to think, and to like to think, that the whole phenomena of life will one day be reduced to terms of matter and motion, and that every vegetable, animal, and human product will be explained, and may some day be artificially produced, by chemical action. But even if this were so, behind it all there would still remain an unexplained mystery."

Closely associated with "Man's Place in the Universe" is a small volume, "Is Mars Habitable?" This was first commenced as a review of Professor Percival Lowell's book, "Mars and its Canals," with the object of showing that the large amount of new and interesting facts contained in this work did not invalidate the conclusion that he (Wallace) had reached in 1903—that Mars is not habitable. The conclusions to which his argument led him were these:

(1) All physicists are agreed that ... Mars would have a mean temperature of about 35° F. owing to its distance from the sun.

(2) But the very low temperatures on the earth under the equator at a height where the barometer stands at about three times as high as on Mars, proves that from scantiness of atmosphere alone Mars cannot possibly have a temperature as high as the freezing-point of water. The combination of these two results must bring down the temperature of Mars to a degree wholly incompatible with the existence of animal life.

(3) The quite independent proof that water-vapour cannot exist on Mars, and that, therefore, the first essential of organic life—water—is non-existent.

The conclusion from these three independent proofs ... is therefore irresistible—that animal life, especially in its highest forms, cannot exist. Mars, therefore, is not only uninhabited by intelligent beings ... but is absolutely uninhabitable.

In contrast to his purely scientific interest in astronomy, Wallace was moved by the romance of the "stars," akin to his enthusiastic love of beautiful butterflies. Had it not been for this touch of romance and idealism in his writings on astronomy, they would have lost much of their charm for the general reader. His breadth of vision transforms him from a mere student of astronomy into a seer who became ever more deeply conscious of the mystery both "before and behind."

"Rain, sun, and rain! and the free blossom blows;

Sun, rain, and sun! and where is he who knows?

From the great deep to the great deep he goes."

And whilst facing with brave and steady mind the great mysteries of earth and sky, of life and what lies beyond it, he himself loved to quote:

"Fear not thou the hidden purpose

Of that Power which alone is great,

Nor the myriad world His shadow,

Nor the silent Opener of the Gate."

Among the scientific friends to whom he appealed for help when writing his astronomical books was Prof. (now Sir) W.F. Barrett.

## TO PROF. BARRETT

*Parkstone, Dorset. February 12, 1901.*

My dear Barrett,—I shall be much obliged if you will give me your opinion on a problem in physics that I cannot find answered in any book. It relates to the old Nebular Hypothesis, and is this:

It is assumed that the matter of the solar system was once wholly gaseous, and extended as a roughly globular or lenticular mass beyond the orbit of Neptune. Sir Robert Ball stated in a lecture here that even when the solar nebula had shrunk to the size of the earth's orbit it must have been (I think he said) hundreds of times rarer than the residual gas in one of Crookes's high vacuum tubes. Yet, by hypothesis, it was hot enough, even in its outer portions, to retain all the solid elements in the gaseous state.

Now, admitting this to be *possible* at any given epoch, my difficulty is this: how long could the outer parts of this nebula exist, exposed to the zero temperature of surrounding space, without losing the gaseous state and aggregating into minute solid particles—into meteoric dust, in fact?

Could it exist an hour? a day? a year? a century? Yet the process of condensation from the Neptunian era to that of Saturn or Jupiter must surely have occupied millions of centuries. What kept the almost infinitely rare metallic gases in the gaseous state all this time? Is such a condition of things physically possible?

I cannot myself imagine any such condition of things as the supposed primitive solar nebula as possibly coming into existence under any conceivably antecedent conditions, but, granted that it did come into existence, it seems to me that the gaseous state must almost instantly begin changing into the solid state. Hence I adopt the meteoric theory instead of the nebular; since all the evidence is in favour of solid matter being abundant all through known space, while there is no evidence of metallic gases existing in space, except as the result of collisions of huge masses of matter. Is my difficulty a mare's nest?—Yours very truly,

ALFRED R. WALLACE.


## TO Mrs. Fisher

*Broadstone, Wimborne. February 28, 1905.*

Dear Mrs. Fisher,—Thanks for your letter. Am sorry I have not converted you, but perhaps it will come yet! I will only make one remark as to your conclusion.

I have not attempted to prove a negative! That is not necessary. What I claim to have done is, to have shown that all the evidence we have, be it much or little, is decidedly against not only other solar planets having inhabitants, but also, as far as probabilities are concerned, equally against it in any supposed stellar planets—for not one has been proved to exist. There is absolutely no evidence which shows even a probability of there being other inhabited worlds. It is all pure speculation, depending upon our ideas as to what the universe is for, as to what *we* think (some of us!) *ought* to be! That is not evidence, even of the flimsiest. All I maintain is that mine *is* evidence, founded on physical probabilities, and that, as against no evidence at all—no proved physical probability—mine holds the field!—Yours very truly,

ALFRED R. WALLACE.

## TO MR. E. SMEDLEY

*Broadstone, Dorset. July 24, 1907.*

Dear Mr. Smedley,— ... I write chiefly to tell you that I have read Mr. Lowell's last book, "Mars and its Canals," and am now writing an article, or perhaps a small book, about it. I am sure his theories are all wrong, and I am showing why, so that anyone can see his fallacies. His observations, drawings, photographs, etc., are all quite right, and I believe true to nature, but his interpretation of what he sees is wrong—often even to absurdity. He began by thinking the straight lines are works of art, and as he finds more and more of these straight lines, he thinks that proves more completely that they are works of art, and then he twists all other evidence to suit that. The book is not very well written, but no doubt the newspaper men think that as he is such a great astronomer he must know what it all means!

I am more than ever convinced that Mars is totally uninhabitable....—Yours very truly,

ALFRED R. WALLACE.

## TO PROF. BARRETT

*Broadstone, Wimborne. August 10, 1907.*

My dear Barrett,—Thanks for your letter, and your friend Prof. Stroud's. I have come to the sad conclusion that it is hopeless to get any mathematician to trouble himself to track out Lowell's obscurities and fallacies.... So, being driven on to my own resources, I have worked out a mode of estimating (within limits) the temperature of Mars, without any mathematical formulæ—and only a little arithmetic. I want to know if there is any fallacy in it, and therefore take the liberty of sending it to you, as you are taking your holiday, just to read it over and tell me if you see any flaw in it. I also send my short summary of Lowell's *Philosophical Magazine* paper, so that you can see if my criticism at the end is fair, and whether his words really mean what to me they seem to....—Yours very sincerely,

ALFRED R. WALLACE.


## TO MR. F. BIRCH

*Sept. 12, 1907.*

Dear Fred,— ... For the last two or three months I have had a hard struggle with Mars—not the god of war, but the planet—writing a small book, chiefly criticising Lowell's last book, called "Mars and its Canals," published less than a year back by Macmillan, who will also publish my reply. *I* think it is crushing, but it has cost me a deal of trouble, as Lowell has also printed a long and complex mathematical article trying to prove that though Mars receives less than half the sun-heat we do, yet it is very nearly as warm and quite habitable! But his figures and arguments are alike so shaky and involved that I cannot get any of my mathematical friends to tackle it or point out his errors. However, I think I have done it myself by the rules of common sense....—Your sincere friend,

ALFRED R. WALLACE.


## TO MR. H. JAMYN BROOKE

*Old Orchard, Broadstone, Wimborne. December 2, 1910.*

Dear Sir,—Your "monistic" system is to me a system of mere contradictory words. You begin with three things—then you say they are correlated with one substance—coextensive with the universe. This you cannot possibly know, and it is about as intelligible and as likely to be true as the Athanasian Creed!—Yours truly,

ALFRED R. WALLACE.

## TO PROF. KNIGHT

*Old Orchard, Broadstone, Dorset. October 1, 1913.*

Dear Mr. Knight,—I have written hardly anything on the direct proofs of "immortality" except in my book on "Miracles and Modern Spiritualism," and also in "My Life," Vol. II. But my two works, "Man's Place in the Universe" (now published at 1s.), and my later volume, "The World of Life," form together a very elaborate, and I think conclusive, scientific argument in favour of the view that the whole material universe exists and is designed for the production of immortal spirits, in the greatest possible diversity of nature, and character, corresponding with ... the almost infinite diversity of that universe, in all its parts and in every detail....—Yours very truly,

ALFRED R. WALLACE.

P.S.—I am fairly well, but almost past work.—A.R.W.

## TO SIR OLIVER LODGE

*Old Orchard, Broadstone, Dorset. October 9, 1913.*

Dear Sir Oliver Lodge,—Owing to ill-health and other causes I have only now been able to finish the perusal of your intensely interesting and instructive Address to the British Association. I cannot, however, refrain from writing to you to express my admiration of it, and especially of the first half of it, in which you discuss the almost infinite variety and complexity of the physical problems involved in the great principle of "continuity" in so clear a manner that outsiders like myself are able to some extent to apprehend them. I am especially pleased to find that you uphold the actual existence and *continuity* of the ether as scientifically established, and reject the doubts of some mathematicians as to the reality and perfect continuity of space and time as unthinkable.

The latter part of the Address is even more important, and is especially notable for your clear and positive statements as to the evidence in all life-process of a "guiding" Mind. I can hardly suppose that you can have found time to read my rather discursive and laboured volume on "The World of Life," written mainly for the purpose of enforcing not only the proofs of a "guiding" but also of a "foreseeing" and "designing" Mind by evidence which will be thought by most men of science to be unduly strained. It is, therefore, the more interesting to me to find that you have yourself (on pp. 33-34 of your Address) used the very same form of analogical illustration as I have done (at p. 296 of "The World of Life") under the heading of "A Physiological Allegory," as being a very close representation of what really occurs in nature.

To conclude: your last paragraph rises to a height of grandeur and eloquence to which I cannot attain, but which excites my highest admiration.

Should you have a separate copy to spare of your Romanes Lecture at Oxford, I should be glad to have it to refer to.—Believe me yours very truly,

ALFRED R. WALLACE.

The last of Wallace's letters on astronomical subjects was written to Sir Oliver Lodge about a week before his death:

## TO SIR OLIVER LODGES

*Old Orchard, Broadstone, Dorset. October 27, 1913.*

Dear Sir Oliver Lodge,—Many thanks for your Romanes Lecture, which, owing to my ignorance of modern electrical theory and experiments, is more difficult for me than was your British Association Address.

I have been very much interested the last month by reading a book sent me from America by Mr. W.L. Webb, being "An Account of the Unparalleled Discoveries of Mr. T.J.J. See."

Several of Mr. See's own lectures are given, with references to his "Researches on the Evolution of the Stellar Systems," in two large volumes.

His theory of "capture" of suns, planets, and satellites seems to me very beautifully worked out under the influence of gravitation and a resisting medium of cosmical dust—which explains the origin and motions of the moon as well as that of all the planets and satellites far better than Sir G. Darwin's expulsion theory.

I note however that he is quite ignorant that Proctor, forty years ago, gave full reasons for this "capture" theory in his "Expanse of Heaven," and also that the same writer showed that the Milky Way could not have the enormous lateral extension he gives to it, but that it cannot really be much flattened. He does not even mention the proofs given of this both by Proctor and, I think, by Herbert Spencer, while in Mr. Webb's volume (opposite p. 212) is a diagram showing the "Coal Sack" as a "vacant lane" running quite through and across the successive spiral extensions laterally of the galaxy, without any reference or a word of explanation that such features, of which there are many, really demonstrate the untenability of such extension.

An even more original and extremely interesting part of Mr. See's work is his very satisfactory solution of the hitherto unsolved geological problem of the origin of all the great mountain ranges of the world, in Chapters X., XI., and

XII. of Mr. Webb's volume. It seems quite complete except for the beginnings, but I suppose it is a result of the formation of the *earth* by accretion and not by expulsion, by heating and not by cooling....—Yours very truly,

D R. WALLACE.

## II.—SPIRITUALISM

"The completely materialistic mind of my youth and early manhood has been slowly moulded into the socialistic, spiritualistic, and theistic mind I now exhibit—a mind which is, as my scientific friends think, so weak and credulous in its declining years, as to believe that fruit and flowers, domestic animals, glorious birds and insects, wool, cotton, sugar and rubber, metals and gems, were all foreseen and foreordained for the education and enjoyment of man. The whole cumulative argument of my 'World of Life' is that *in its every detail* it calls for the agency of a mind ... enormously above and beyond any human mind ... Whether this Unknown Reality is a single Being and acts everywhere in the universe as direct creator, organiser, and director of every minutest motion ... or through 'infinite grades of beings,' as I suggest, comes to much the same thing. Mine seems a more clear and intelligible supposition ... and it is the teaching of the Bible, of Swedenborg, and of Milton."—Letter from A.R. Wallace to JAMES MARCHANT, written in 1913.

The letters on Spiritualism which Wallace wrote cast further light on the personal attitude of mind which he maintained towards that subject. He was an unbiased scientific investigator, commencing on the "lower level" of spirit phenomena, such as raps and similar physical manifestations of "force by unseen intelligences," and passing on to a clearer understanding of the phenomena of mesmerism and telepathy; to the materialisation of, and conversation with, the spirits of those who had been known in the body, until the conviction of life after death, as the inevitable crowning conclusion to the long process of evolution, was reached in the remarkable chapter with which he concludes "The World of Life"—an impressive prose poem.

Like that of many other children, Wallace's early childhood was spent in an orthodox religious atmosphere, which, whilst awakening within him vague emotions of religious fervour, derived chiefly from the more picturesque and impassioned of the hymns which he occasionally heard sung at a Nonconformist chapel, left no enduring impression. Moreover, at the age of 14 he was brought suddenly into close contact with Socialism as expounded by Robert Owen, which dispelled whatever glimmerings of the Christian faith there may have been latent in his mind, leaving him for many years a confirmed materialist.

This fact, together with his early-aroused sense of the social injustice and privations imposed upon the poorer classes both in town and country, which he carefully observed during his experience as a land-surveyor, might easily have had an undesirable effect upon his general character had not his intense love and reverence for nature provided a stimulus to his moral and spiritual development. But the "directive Mind and Purpose" was preparing him silently and unconsciously until his "fabric of thought" was ready to receive spiritual impressions. For, according to his own theory, as "the laws of nature bring about continuous development, on the whole progressive, one of the subsidiary results of this mode of development is that no organ, no sensation, no faculty arises *before* it is needed, or in greater degree than it is needed."[57] From this point of view we may make a brief outline of the manner in which this particular "faculty" arose and was developed in him.

When at Leicester, in 1844, his curiosity was greatly excited by some lectures on mesmerism given by Mr. Spencer Hall, and he soon discovered that he himself had considerable power in this direction, which he exercised on some of his pupils.

Later, when his brother Herbert joined him in South America, he found that he also possessed this gift, and on several occasions they mesmerised some of the natives for mere amusement. But the subject was put aside, and Wallace paid no further attention to such phenomena until after his return to England in 1862.

It was not until the summer of 1865 that he witnessed any phenomena of a spiritualistic nature; of these a full account is given in "Miracles and Modern Spiritualism" (p. 132). "I came," he says, "to the inquiry utterly unbiased by hopes or fears, because I knew that my belief could not affect the reality, and with an ingrained prejudice even against such a word as 'spirit,' which I have hardly yet overcome."

From that time until 1895, when the second edition of that book appeared, he did much, together with other scientists, to establish these facts, as he believed them to be, on a rational and scientific foundation. It will also be noticed, both before and after this period, that in addition to the notable book which he published dealing exclusively with these matters, the gradual trend of his convictions, advancing steadily towards the end which he ultimately reached, had become so thoroughly woven into his "fabric of thought" that it appears under many phases in his writings, and occupies a considerable part of his correspondence, of which we have only room for some specimens.

The first definite statement of his belief in "this something" other than material in the evolution of Man appeared in his essay on "The Development of Human Faces under the Law of Natural Selection" (1864). In this he suggested that, Man having reached a state of physical perfection through the progressive law of Natural Selection, thenceforth Mind became the dominating factor, endowing Man with an ever-increasing power of intelligence which, whilst the physical had remained stationary, had continued to develop according to his needs. This "in-breathing" of a divine Spirit, or the controlling force of a supreme directive Mind and Purpose, which was one of the points of divergence between his theory and that held by Darwin, is too well known to need repetition.

This disagreement has a twofold interest from the fact that Darwin, in his youth, studied theology with the full intention of taking holy orders, and for some years retained his faith in the more or less orthodox beliefs arising out of the Bible. But as time went by, an ever-extending knowledge of the mystery of the natural laws governing the development of man and nature led him to make the characteristically frank avowal that he "found it more and more difficult ... to invent evidence which would suffice to convince"; adding, "This disbelief crept over me at a very slow rate, but was at last complete. The rate was so slow that I felt no distress."[58] With Wallace, however, his early disbelief ended in a deep conviction that "as nothing in nature actually 'dies,' but renews its life in another and higher form, so Man, the highest product of natural laws here, must by the power of mind and intellect continue to develop hereafter."

The varied reasons leading up to this final conviction, as related by himself in "Miracles and Modern Spiritualism" and "My Life," are, however, too numerous and detailed to be retold in a brief summary in this place.

The correspondence that follows deals entirely with investigations on this side of the Atlantic, but a good deal of evidence which to him was conclusive was obtained during his stay in America, where Spiritualism has been more widely recognised, and for a much longer period than in England.

Some of the letters addressed to Miss Buckley (afterwards Mrs. Fisher) reveal the extreme caution which he both practised himself and advocated in others when following up any experimental phase of spiritual phenomena. The same correspondence also gives a fairly clear outline of his faith in the ascending scale from the physical evidence of spirit-existence to the communication of some actual knowledge of life as it exists beyond the veil.

In spiritual matters, as in natural science, though at times his head may have appeared to be "in the clouds," his feet were planted firmly on the earth. This is seen, to note another curious instance, in his correspondence with Sir Wm. Barrett, where he maintains a delicate balance between natural science and "spirit impression" when discussing the much controverted reality of "dowsing" for water.

It was this breadth of vision, unhampered by mere intellectualism, but always kept within reasonable bounds by scientific deduction and analysis, which constituted Alfred Russel Wallace a seer of the first rank.

Wallace lived to see the theory of evolution applied to the life-history of the earth and the starry firmament, to the development of nations and races, to the progress of mind, morals and religion, even to the origin of consciousness and life—a conception which has completely revolutionised man's attitude towards himself and the world and God. Evolution became intelligible in the light of that idea which came to him in his hut at Ternate and changed the face of the universe. Surely it was enough for any one man to be one of the two chief originators of such a far-reaching thought and to witness its impact upon the ancient story of special creations which it finally laid in the dust. But Wallace was privileged beyond all the men of his generation. He lived to see many of the results of the theory of evolution tested by time and to foresee that there were definite limits to its range, that, indeed, there were two lines of development—one affecting the visible world of form and colour and the other the invisible world of life and spirit—two worlds springing from two opposite poles of being and developing *pari passu*, or, rather, the spiritual dominating the material, life originating and controlling organisation. It was, in short, his peculiar task to reveal something of the Why as well as the How of the evolutionary process, and in doing so verily to bring immortality to light.

The immediate exciting cause of this discovery of the inadequacy of evolution from the material side alone to account for the world of life may seem to many to have been trivial and unworthy of the serious attention of a great scientist. How, it might be asked, could the crude and doubtful phenomena of Spiritualism afford reasonably adequate grounds for challenging its supremacy and for setting a limit to its range? But spiritualistic phenomena were only the accidental modes in which the other side of evolution struck in upon his vision. They set him upon the other track and opened up to him the vaster kingdom of life which is without beginning, limit or end; in which perchance the sequence of life from the simple to the complex, from living germ to living God, may also be the law of growth. It is in the light of this ultimate end that we must judge the stumbling steps guided by raps and visions which led him to the ladder set up to the stars by which connection was established with the inner reality of being. That was the distinctive contribution which he made to human beliefs over and above his advocacy of pure Darwinism.

Reading almost everything he could obtain upon occult phenomena, Wallace found that there was such a mass of testimony by men of the highest character and ability in every department of human learning that he thought it would be useful to bring this together in a connected sketch of the whole subject. This he did, and sent it to a secularist magazine, in which it appeared in 1866, under the title of "The Scientific Aspect of the Supernatural." He sent a copy to Huxley.

### TO T.H. HUXLEY

*9 St. Mark's Crescent, Regent's Park, N.W. November 22, 1866.*

Dear Huxley,—I have been writing a little on a *new branch* of Anthropology, and as I have taken your name in vain on the title-page I send you a copy. I fear you will be much shocked, but I can't help it; and before finally deciding that we are all mad I hope you will come and see some very curious phenomena which we can show you, *among friends only*. We meet every Friday evening, and hope you will come sometimes, as we wish for the fullest investigation, and shall be only too grateful to you or anyone else who will show us how and where we are deceived.

### T.H. HUXLEY TO A.R. WALLACE

[? *November, 1886.*]

Dear Wallace,—I am neither shocked nor disposed to issue a Commission of Lunacy against you. It may be all true, for anything I know to the contrary, but really I cannot get up any interest in the subject. I never cared for gossip in my life, and disembodied gossip, such as these worthy ghosts supply their friends with, is not more interesting to me than any other. As for investigating the matter, I have half-a-dozen investigations of infinitely greater interest to me to which any spare time I may have will be devoted. I give it up for the same reason I abstain from chess—it's too amusing to be fair work, and too hard work to be amusing.—Yours faithfully,

T.H. HUXLEY.

## TO T.H. HUXLEY

*9 St. Mark's Crescent, Regent's Park, N.W. December 1, 1866.*

Dear Huxley,—Thanks for your note. Of course, I have no wish to press on you an inquiry for which you have neither time nor inclination. As for the "gossip" you speak of, I care for it as little as you can do, but what I do feel an intense interest in is the exhibition of *force* where force has been declared *impossible*, and of *intelligence* from a source the very mention of which has been deemed an *absurdity*.

Faraday has declared (apropos of this subject) that he who can prove the existence or exertion of force, if but the lifting of a single ounce, by a power not yet recognised by science, will deserve and assuredly receive applause and gratitude. (I quote from memory the sense of his expressions in his Lecture on Education.)

I believe I can now show such a force, and I trust some of the physicists may be found to admit its importance and examine into it.—Believe me yours very sincerely,

ALFRED R. WALLACE.

## TO MISS BUCKLEY

*Holly House, Barking, E. December 25, 1870.*

Dear Miss Buckley,— ... You did not hear Mrs. Hardinge[52] on very favourable topics, and I hope you will hear her often again, and especially hear one of her regular discourses. I think, however, from what you heard, that, setting aside all idea of her being more than a mere spiritualist lecturer setting forth the ideas and opinions of the sect, you will admit that spiritualists, as represented by her, are neither prejudiced nor unreasonable, and that they are truly imbued with the scientific spirit of subordinating all theory to fact. You will also admit, I think, that the moral teachings of Spiritualism, as far as she touched upon them, are elevated and beautiful and calculated to do good; and if so, that is the use of Spiritualism—the getting such doctrines of future progress founded on actual phenomena which we can observe and examine now, not on phenomena which are said to have occurred thousands of years ago and of which we have confessedly but imperfect records.

I think, too, that the becoming acquainted with two such phases of Spiritualism as are exhibited by Mrs. Hardinge and Miss Houghton must show you that the whole thing is not to be judged by the common phenomena of public stances alone, and I can assure you that there are dozens of other phases of the subject as remarkable as these two....—Yours very faithfully,

ALFRED R. WALLACE.

## TO MISS BUCKLEY

*Holly House. Barking, E. June 1, 1871.*

Dear Miss Buckley,— ... I have lately had a stance with the celebrated Mr. Home, and saw that most wonderful phenomenon an accordion playing beautiful music by itself, the bottom only being held in Mr. Home's hand. I was invited to watch it as closely as I pleased under the table in a well-lighted room. I am sure nothing touched it but Mr. Home's one hand, yet at one time I saw a shadowy yet defined hand on the keys. This is too vast a phenomenon for any sceptic to assimilate, and I can well understand the impossibility of their accepting the evidence of their own senses. Mr. Crookes, F.R.S., the chemist, was present and suspended the table with a spring balance, when it was at request made heavy or light, the indicator moving accordingly, and to prevent any mistake it was made light when the hands of all present were resting on the table and heavy when our hands were all underneath it. The difference, if I remember, was about 40 lb. I was also asked to place a candle on the floor and look under the table while it was lifted completely off the floor, Mr. Home's feet being 2 ft. distant from any part of it. This was in a lady's house in the West End. Mr. Home courts examination if people come to him in a fair and candid spirit of inquiry....— Yours very faithfully,

ALFRED R. WALLACE.

## TO MISS BUCKLEY

*The Dell, Grays, Essex. January 11, 1874.*

My dear Miss Buckley,—I am delighted to hear of your success so far, and hope you are progressing satisfactorily. Pray keep accurate notes of all that takes place.... Allow me ... to warn you not to take it for granted till you get proof upon proof that it is really your sister that is communicating with you.

I hope and think it is, but still, the conditions that render communication possible are so subtle and complex that she may not be able; and some other being, reading your mind, may be acting through you and making you think it is your sister, to induce you to go on. Be therefore on the look out for characteristic traits of your sister's mind and manner which are different from your own. These will be tests, especially if they come when and how you are not expecting them. Even if it is your sister, she may be obliged to use the intermediation of some other being, and in that case her peculiar idiosyncrasy may be at first disguised, but it will soon make itself distinctly visible. Of course you will preserve every scrap you write, and date them, and they will, I have no doubt, explain each other as you go on.

If you can get to see the last number of the *Quarterly Journal of Science*, you will find a most important article by Mr. Crookes, giving an outline of the results of his investigations, which he is going to give in full in a volume. His facts are most marvellous and convincing, and appear to me to answer every one of the objections that have usually been made to the evidence adduced....—Yours very faithfully,

ALFRED R. WALLACE.


## TO MISS BUCKLEY

*The Dell, Grays, Essex. February 28, 1874.*

Dear Miss Buckley,—I was much pleased with your long and interesting letter of the 19th and am glad you are getting on at last. It will be splendid if you really become a good medium for some first-rate unmistakable manifestations that even Huxley will acknowledge are worth seeing, and Carpenter confess are not to be explained by unconscious cerebration....—Yours very faithfully,

ALFRED R. WALLACE.


## TO MISS BUCKLEY

*The Dell, Grays, Essex. March 9, 1874.*

Dear Miss Buckley,—I compassionate your mediumistic troubles, but I have no doubt it will all come right in the end. The fact that your sister will not talk as you want her to talk—will not say what you expect her to say, is a grand proof that it is not your unconscious cerebration that does her talking for her. Is not that clear? Whether it is she herself or someone else who is talking to you, is not so clear, but that it is not you, I think, is clear enough.

I can quite understand, too, that your sister in her new life may be, above all things, interested in getting the telegraph in good order, to communicate, and will not think of much else till that is done. While the first Atlantic cable was being laid the messages would be chiefly reports of progress, directions and instructions, with now and then trivialities about the weather, the time, or small items of news. Only when it was in real working order was a President's Message, a Queen's Speech, sent through it.

Automatic writing and trance speaking never yet convinced anybody. They are only useful for those who are already convinced. But you *would* begin this way. You would not go to mediums and séances and see what you could get that way. So now you must persevere; but do not give up your own judgment in anything. Insist upon having things explained to you, or say you won't go on. You will then find they will be explained, only it may take a little more time.... —Yours very faithfully,

ALFRED R. WALLACE.

## TO MISS BUCKLEY

*The Dell, Grays, Essex. April 24, 1874.*

Dear Miss Buckley,— ... On coming home this evening I received the news of poor little Bertie's death—this morning at eight o'clock. I left him only yesterday forenoon, and had then considerable hopes, for we had just commenced a new treatment which a fortnight earlier I am pretty sure might have saved him. The thought suddenly struck me to go to Dr. Williams, of Hayward's Heath ... but it was too late. As he had been in this same state of exhaustion for nearly a month, it is evident that very slight influences might have been injurious or beneficial. Our orthodox medical men are profoundly ignorant of the subtle influences of the human body in health and disease, and can thus do nothing in many cases which Nature would cure if assisted by proper conditions. We who know what strange and subtle influences are around us can believe this....—Yours very truly,

ALFRED R. WALLACE.

Mr. Wallace felt the death of this child so deeply that during the remainder of his life he never mentioned him except when obliged, and then with tears in his eyes.—A.B. FISHER.

## TO MISS BUCKLEY

*The Dell, Grays, Essex. Thursday evening, [? December, 1875].*

Dear Miss Buckley,—Our stance came off last evening, and was a tolerable success. The medium is a very pretty little lively girl, the place where she sits a bare empty cupboard formed by a frame and doors to close up a recess by the side of a fireplace in a small basement breakfast-room. We examined it, and it is absolutely impossible to conceal a scrap of paper in it. Miss Cooke is locked in this cupboard, above the door of which is a square opening about 15 inches each way, the only thing she takes with her being a long piece of tape and a chair to sit on. After a few minutes Katie's whispering voice was heard, and a little while after we were asked to open the door and seal up the medium. We found her hands tied together with the tape passed three times round each wrist and tightly knotted, the hands tied close together, the tape then passing behind and well knotted to the chair-back. We sealed all the knots with a private seal of my friend's, and again locked the door. A portable gas lamp was on a table the whole evening, shaded by a screen so as to cast a shadow on the square opening above the door of the cupboard till permission was given to illuminate it. Every object and person in the room were always distinctly visible. A face[60] then appeared at the opening, but dark and indistinct.

After a time another face quite distinct with a white turban-like headdress—this was a handsome face with a considerable general likeness to that of the medium, but paler, larger, fuller, and older—decidedly a different face, although like. The light was thrown full on this face, and on request it advanced so that the chin projected a little beyond the aperture. We were then ordered to release the medium. I opened the door, and found her bent forward with her head in her lap, and apparently in a deep sleep or trance—from which a touch and a few words awoke her. We then examined the tape and knots—all was as we left it and every seal perfect.

The same face appeared later in the evening, and also one decidedly different with coarser features.

After this, for the sake I believe of two sceptics present, the medium was twice tied up in a way that no human being could possibly tie herself. Her wrists were tied together so tightly and painfully that it was impossible to untie them in any moderate time, and she was also secured to the chair; on the other occasion the two arms were tied close above the elbows so tightly that the arms were swelling considerably from impeded circulation, the elbows being drawn together as close as possible behind the back, there repeatedly knotted, and again tightly knotted to the back of the chair. Miss

C. was evidently in considerable pain, and she had to be lifted out bodily in her chair before we could safely cut her loose, so tightly was she bound. This evidently had a great effect on the sceptics, as I have no doubt it was intended to have, and it demonstrated pretty clearly that some strange being was inside the cupboard playing these tricks, although quite invisible and intangible to us except when she made certain portions of herself visible.

When Miss C. was complaining of being hurt by the tying we could hear the whispering voice soothing her in the kindest manner, and also heard kisses, and Miss C. afterwards declared that she could feel hands and face about her like those of a real person.

During all the face exhibitions singing had to go on to a rather painful extent.[61]

A Dr. Purdon was present, an Army surgeon, who has been much in India, and seems a very intelligent man. He seemed very intimate with the family, and told us he had studied them all, and had had Miss Cooke a month at a time in his own house, studying these phenomena. He was absolutely satisfied of their genuineness, and indeed no opportunity for imposture seems to exist.

The children of the house tell wonderful tales of how they are lifted up and carried about by the spirits. They seem to enjoy it very much, and to look upon it all as just as real and natural as any other matters of their daily life.

Can such things be in this nineteenth century, and the wise ones pass away in utter ignorance of their existence?—Yours very sincerely,

ALFRED R. WALLACE.

At the Glasgow Meeting of the British Association in 1876, Prof. (now Sir) W.F. Barrett read a paper "On some Phenomena associated with Abnormal Conditions of Mind." Wallace was Chairman of the Section in which the paper was read, and a vigorous controversy arose at the close between Dr. Carpenter, who came in towards the end of the paper, and the Chairman. The paper set forth certain remarkable evidence which Prof. Barrett had obtained from a subject in the mesmeric trance, giving what appeared to be indubitable proof of some supernormal mode of transmission of ideas from his mind to that of the subject. The facts were so novel and startling that Prof. Barrett asked for a committee of experts to examine the whole question and see whether such a thing as "thought transference," independently of the recognised channels of sense, did really exist. This was the first time evidence of this kind had been brought before a scientific society, and a protracted discussion followed. The paper also dealt with certain so-called spiritualistic phenomena, which at the time Prof. Barrett was disposed to attribute to hallucination and "thought-transference." The introduction of this topic led the discussion away from the substance of the paper, and Prof. Barrett's plea for a committee of investigation on thought-transference fell through. So strong was the feeling against the paper in official scientific circles at the time, that even an abstract was refused publication in the *Report* of the British Association, and it was not until the Society for Psychical Research was founded that the paper was published, in the first volume of its *Proceedings*. It was the need of a scientific society to collect, sift and discuss and publish the evidence on behalf of such supernormal phenomena as Prof. Barrett described at the British Association that induced him to call a conference in London at the close of 1881, which led to the foundation of the Society for Psychical Research early in 1882.

Wallace, in his letter to Prof. Barrett which follows, refers to Reichenbach's experiments with certain sensitives who declared they saw luminosity from the poles of a magnet after they had been for some time in a perfectly darkened room. Acting on Wallace's suggestion, Prof. Barrett constructed a perfectly darkened room and employed a large electro-magnet, the current for which could be made or broken by an assistant outside without the knowledge of those present in the darkened room. Under these circumstances, and taking every precaution to prevent any knowledge of when the magnet was made active by the current, Prof. Barrett found that two or three persons, out of a large number with whom he experimented, saw a luminosity streaming from the poles of the magnet directly the current was put on. An article of Prof. Barrett's on the subject, with the details of the experiment, was published in the *Philosophical Magazine*, and also in the *Proceedings* of the Society for Psychical Research (Vol. I.).

## TO PROF. BARRETT

*Rosehill, Dorking, December 18, 1876.*

My dear Prof. Barrett,— ... I see you are to lecture at South Kensington the end of this month (I think), and if you can spare time to run down here and stay a night or two we shall be much pleased to see you, and I shall be greatly interested to have a talk on the subject of your paper, and hear what further evidence you have obtained. I want particularly to ask you to take advantage of any opportunity that you may have to test the power of sensitives to see the "flames" from magnets and crystals, as also to *feel* the influence from them. This is surely a matter easily tested and settled. I consider it has been tested and settled by Reichenbach, but he is ignored, and a fresh proof of this one fact, by indisputable tests, is much needed; and a paper describing such tests and proofs would I imagine be admitted into the *Proceedings* of any suitable society.

You will have heard no doubt of the Treasury having taken up the prosecution of Slade. Massey the barrister, one of the most intelligent and able of the Spiritualists (whose accession to the cause is due, I am glad to say, to my article in the *Fortnightly*), proposes a memorial and deputation to the Government protesting against this prosecution by the Treasury on the ground that it implies that Slade is an habitual impostor and nothing else, and that in face of the body of evidence to the contrary, it is an uncalled-for interference with the private right of investigation into these subjects. On such general grounds as these I sincerely hope you will give your name to the memorial....—Yours very faithfully,

A.R. WALLACE.

## TO PROF. BARRETT

*Rosehill, Dorking. December 9, 1877.*

My dear Barrett,—I am always glad when a man I like and respect treats me as a friend. I am advised by other friends also not to waste more time on Dr. C. [Carpenter], and I do not think I shall answer him again, except perhaps to keep him to certain points, as in my letter in the last *Nature*. In a proof of his new edition of "Lectures" I see he challenges me to produce a person who can detect by light or sensation when an electro-magnet is made and unmade. The Association of Spiritualists are going to experiment, as Dr. C. offers to pay £30 if it succeeds. Should you have an opportunity of trying with any persons, and can find one who sees or feels the influence strongly, it might be worth while to send him to London, as nothing would tend to lower Dr. C. in public estimation on this subject more than his being forced to acknowledge that what he has for more than thirty years declared to be purely subjective is after all an objective phenomenon.

I never had anything to do with showing or sending a medium to Huxley. He must refer to his séance a few months ago with Mrs. Kane and Mrs. Jencken (along with Carpenter and Tyndall), when ... nothing but raps occurred....—Yours very faithfully,

ALFRED R. WALLACE.

The British Association met in Dublin in 1878, and Prof. Barrett asked Wallace to stay with him at Kingstown, or, if he preferred being nearer the meetings, with a friend in Dublin. Earlier in the year Mr. Huggins, afterwards Sir W. Huggins, O.M. and President of the Royal Society, had sent Prof. Barrett a very beautifully executed drawing of the knots tied in an endless cord during the remarkable sittings Prof. Zöllner had with the medium Slade. Sir W. Huggins invited Prof. Barrett to come and see him at his observatory at Tulse Hill, near London, and there he met Wallace and discussed the whole matter. It may not be generally known that so careful and accurate an observer as Sir W. Huggins was convinced of the genuineness of the phenomena he had witnessed with Lord Dunraven and others through the medium D.D. Home. He informed Prof. Barrett of this himself.

## TO PROF. BARRETT

*Waldron Edge, Duppas Hill, Croydon. June 27, 1873.*

My dear Barrett,—The receipt of a British Association circular reminds me of your kind invitation to stay with you or your friend at Dublin, and as you may be wishing soon to make your arrangements I write at once to let you know that, much to my regret, I shall not be able to come to Dublin this year. Since I met you at Mr. Huggins's I have done nothing myself in Spiritual investigations, but have been exceedingly interested in the knot-tying experiment of Prof. Zöllner and the weight-varying experiments of the Spiritualists' Association. I do not see what flaw can be found in either of them....—Yours very faithfully,

ALFRED R. WALLACE.

In the discussion on Prof. Barrett's paper at the Glasgow Meeting of the British Association, which took place in the London *Times* and other newspapers, instances of apparent thought-transference were given by many correspondents. Each of these cases Prof. Barrett investigated personally, and one of them led to a remarkable series of experiments which he conducted at Buxton, with the result that no doubt was left on his mind of the fact of the transference of ideas from one mind to another independent of the ordinary channels of sense. He asked Prof. and Mrs. H. Sidgwick to come to Buxton and repeat his experiments with the subjects there—daughters of a local clergyman. They did so, and though they had less success at first than Prof. Barrett had had, they were ultimately convinced of the genuineness of the phenomena. In addition, Mr. Edmund Gurney, Mr. Frederic Myers, Prof. A. Hopkinson and Prof. Balfour Stewart, all responded to Prof. Barrett's invitation to visit Buxton and test the matter for themselves, and all came to the same conclusion as he had. Subsequently Gurney and

Myers associated their name with Barrett's in a paper on the subject, published in the *Nineteenth Century*.

Prof. Barrett asked Wallace to read over the first report made by Prof. and Mrs. Sidgwick, which at first seemed somewhat disheartening, and the following is his reply:

## REMARKS ON EXPERIMENTS IN THOUGHT READING BY MR. AND MRS. SIDGWICK AT BUXTON

The failure of so many of these experiments seems to me to depend on their having been conducted without any knowledge of the main peculiarity of thought reading or clairvoyance—that it is a perception of the object thought of or hidden, not by its name, or even by its sum total of distinctive qualities, but by the simple qualities separately. A clairvoyant will perceive a thing as round, then as yellow, and finally as an orange. Now Mr. Galton's experiments have shown how various are the powers of visualising objects possessed by different persons, and how distinct their modes of doing so; and if these distinct visualisations of the same thing are in any way presented to a clairvoyant, there is little wonder that some confusion should result. This would suggest that one person who possesses the faculty of clearly visualising objects would meet with more success than a number of persons some of whom visualise one portion or quality of the object, some another, while to others the name alone is present to the mind. It follows from these considerations that cards are bad for such experiments. The qualities of number, colour, form and arrangement may be severally most prominent in one mind or other, and the result is confusion to the thought reader. This is shown in the experiments by the number of pips or the suit alone being often right.

It must also be remembered that children have not the same thorough knowledge of the names of the cards that we have, nor can they so rapidly and certainly count their numbers. This introduces another source of uncertainty which should be avoided in such experiments as these.

The same thing is still more clearly shown by the way in which objects are guessed by some prominent quality or resemblance, not by any likeness of name—as poker guessed for walking-stick, fork for pipe, something iron for knife, etc. And the total failure in the case of names of towns is clearly explained by the fact that these would convey no distinct idea or concrete image that could be easily described. These last failures really give an important clue to the nature of the faculty that is being investigated, since they show that it is not *words* or *names* that are read but thoughts or images that are perceived, and the certainty of the perception will depend upon the simple character of these images and the clearness and identity of the perception of them by the different persons present.

If these considerations are always kept in view, I feel sure that the experiments will be far more successful.

ALFRED R. WALLACE.

Sept. 6, 1881.

---

Wallace's remarkable gifts as a lecturer are less widely known than his lucid and admirable style as a writer. Though Sir Wm. Barrett has heard a great number of eminent scientific men lecture, he considers that few could approach him for the simplicity, clearness and vigour of his exposition, which commanded the unflagging attention of every one of his hearers. Mr. Frederic Myers, no mean judge of literary merit, once said he thought Wallace one of the most lucid English writers and lecturers of his time. Prof. Barrett was anxious to induce Wallace to lecture in Dublin, and brought the matter before the Science Committee of the Royal Dublin Society, which arranges a course of afternoon lectures by distinguished men every spring. The Committee cordially supported the suggestion that Wallace should be invited to lecture, and the invitation was accepted. During his visit to Dublin, Wallace stayed with Prof. Barrett at Kingstown, and was busily engaged in revising the proof-sheets of his book on "Land Nationalisation" (1882).

In "My Life" (Vol. II., p. 334) Wallace says that among the eminent men whose "first acquaintance and valued friendship" he owed to a common interest in Spiritualism was Frederic Myers, whom he met first at some séances in London about the year 1878.

## F.W.H. MYERS TO A.R. WALLACE

*Leckhampton House, Cambridge. April 12, 1890.*

My dear Wallace,—I will read your pamphlet[62] most carefully; will write and tell you how it affects me; and will in any case send it on with your letter and a letter of my own to Sir John Gorst, whom I know well, and whom I agree with you in regarding as the most acceptable member of the Government.

If I am converted, it will be wholly *your* doing. I have read much on the subject—Creighton, etc., and am at present strongly pro-vaccination; at the same time, there is no one by whom I would more willingly be converted than yourself.

I am glad to take this opportunity of telling you something about my relation to one of your books. I write now from bed, having had some influenzic pneumonia, now going off. For some days my temperature was 105 and I was very restless at night, anxious to read, but in too sensitive and fastidious a state to tolerate almost any book. I found that almost the only book which I could read was your "Malay Archipelago" (of course I had read it before). In spite of my complete ignorance of natural history there was a certain charm about the book, both moral and literary, which made it deeply congenial in those trying hours. You have had few less instructed readers, but very few can have dwelt on that simple manly record with a more profound sympathy.

I want to bespeak you as a *friend at court*. When we get into the next world, I beg you to remember me and say a good word for me when you can, as you will have much influence there.

To me it seems that Hodgson's report[63] is the *best* thing which we have yet published. I trust that it impresses you equally. It has converted *Podmore* amongst other people!

I will, then, write again soon, and I am yours most truly,

F.W.H. MYERS.

## TO MRS. FISHER (*née* BUCKLEY)

*Parkstone, Dorset. January 4, 1896.*

My dear Mrs. Fisher,—I am glad to hear that you are going on with your book. I am sure it will be a comfort to you. I have read one book of Hudson's—"A Scientific Demonstration of a Future Life," and that is so pretentious, so unscientific, and so one-sided that I do not feel inclined to read more of the same author's work. I do not think I mentioned to you (as I thought you did not read much now) a really fine and original work, called "Psychic Philosophy, a Religion of Natural Law," by Desertis (Redway). I should like to know if, after reading that, you still think Hudson's books worth reading.

I have been much pleased and interested lately in reading Mark Twain's, Mrs. Oliphant's and Andrew Lang's books about Joan of Arc. The last two are far the best, Mrs. Oliphant's as a genuine sympathetic *history*, Lang's as a fine realistic story ("A Monk of Fife"). Jeanne was really perhaps the most beautiful character in authentic history, and the one that most conclusively demonstrates spirit-guidance, and both Mrs. Oliphant and A. Lang bring this out admirably.... —Yours very faithfully,

## TO MRS. FISHER

*Parkstone, Dorset. September 14, 1896.*

My dear Mrs. Fisher,—I have much pleasure in signing your application for the Psychical Research Society, though the majority of the active members are so absurdly and illogically sceptical that you will not find much instruction in their sayings. Mr. Podmore's report in the last-issued *Proceedings* is a good illustration....

We have all been in Switzerland this year. Violet, her mother, and five lady friends all went together to a rather newly-discovered place, Adelboden, a branch valley from that going up to the Gemmi Pass by Kandersteg. I went first for a week to Davos, to give a lecture to Dr. Lunn's party, and enjoyed myself much, chiefly owing to the company of Rev. Hugh Price Hughes, one of the most witty, earnest, advanced, and estimable men I have ever met. Dr. Lunn himself is very jolly, and we had also Mr. Le Gallienne, the poet and critic, and between them we had a very brilliant table-talk. Mr. Haweis was also there, and one afternoon he and I talked for two hours about Spiritualism. He is a thorough spiritualist, and preaches it....—Yours very sincerely,

ALFRED R. WALLACE.

## TO MRS. FISHER

*Parkstone, Dorset. April 9, 1897.*

My dear Mrs. Fisher,—I have tried several Reincarnation and Theosophical books, but *cannot* read them or take any interest in them. They are so purely imaginative, and do not seem to me rational. Many people are captivated by it—I think most people who like a grand, strange, complex theory of man and nature, given with authority—people who if religious would be Roman Catholics. Crookes gave a suggestive and interesting, but in some ways rather misleading address as President of the Psychical Research Society. I liked Oliver Lodge's address to the Spiritualists' Association better....—Yours very sincerely,

ALFRED R. WALLACE.

In 1891, at the urgent request of Prof. H. Sidgwick, President of the Society for Psychical Research, Prof. Barrett undertook, with considerable reluctance, to make a thorough examination of the subject of "dowsing" for water and minerals by means of the so-called "divining rod." At the time he fully believed that a critical inquiry of this kind would speedily show all the alleged successes of the dowser to be due either to fraud or a sharp eye for the ground. As the inquiry went on, to his surprise he found that neither chicanery, nor clever guessing, nor local knowledge, nor chance coincidence could explain away the accumulated evidence, but that something new to science was really at the root of the matter. This result was so startling that Prof. Barrett had to pursue the investigation for six years before venturing to publish his first report, which appeared in the *Proceedings* of the Society for Psychical Research, Part xxxii., 1897. This was followed by a second report published some years later, in which he gave a fresh body of evidence on the criticisms of some eminent geologists to whom he had submitted the evidence. The reports were reviewed in *Nature* with considerable severity, and some erroneous statements were made, to which Prof. Barrett replied. The editor, Sir Norman Lockyer, at first declined to publish Prof. Barrett's reply, and to this Wallace refers in the following letter.

### TO PROF. BARRETT

*Parkstone, Dorset. October 30, 1899.*

My dear Barrett,— ... Apropos of *Nature*, they never gave a word of notice to my book[64]—probably they would say out of kindness to myself as one of their oldest contributors, since they would have had to scarify me, especially as regards the huge Vaccination chapter, which is nevertheless about the most demonstrative bit of work I have done. I begged Myers—as a personal favour—to read it. He told me he firmly believed in vaccination, but would do so, and afterwards wrote me that he could see no answer to it, and if there was none he was converted. There certainly has been not a tittle of answer except abuse.

I am glad you brought Lockyer up sharp in his attempt to refuse you the right to reply. I am glad you now have some personal observations to adduce. I hope persons or corporations who are going to employ a dowser will now advise you so that you may be present....—Yours very faithfully,

ALFRED R. WALLACE.

## TO PROF. BARRETT

*Parkstone, Dorset. December 24, 1900.*

My dear Barrett,— ... I have read your very interesting paper on the divining rod, and the additional evidence you now send. Of course, I think it absolutely conclusive, but there are many points on which I differ from your conclusions and remarks, which I think are often unfair to the dowsers.

I will just refer to one or two. At p. 176 (note) you call the idea of there being a "spring-head" at a particular point "absurd." But instead of being absurd it is a *fact*, proved not only by numerous cases you have given of strong springs being found quite near to weak springs a few yards off, but by all the phenomena of mineral and hot springs. Near together, as at Bath, hot springs and cold springs rise to the surface, and springs of different quality at Harrogate, yet each keeps its distinct character, showing that each rises from a great depth without any lateral diffusion or intermixture. This is a common phenomenon all over the world, the dowsers' facts support it, geologists know all about it, yet I presume they have told you that when a dowser states this fact it ceases to be a fact and becomes an absurdity!

The only other point I have time to notice is your Sect. II. (p. 285). You head this, "Evidence that the Motion of the Rod is due to Unconscious Muscular Action." Naturally I read this with the greatest interest, but found to my astonishment that you adduce no evidence at all, but only opinions of various people, and positive assertions that such is the case! Now as I *know* that motions of various objects occur without any muscular action, or even any contact whatever, while Crookes has proved this by careful experiments which have never been refuted, what *improbability* is there that this should be such a case, and what is the value of these positive assertions which you quote as "evidence"? And at p. 286 you quote the person who says the more he tried to prevent the stick's turning the more it turned, as *evidence* in favour of muscular action, without a word of explanation. Another man (p. 287) says he "could not restrain it." None of the "trained anatomists" you quote give a particle of *proof*, only positive opinion, that it must be muscular action— simply because they do not believe any other action possible. Their evidence is just as valueless as that of the people who say that all thought-transference is collusion or imposture!

I do not say that it is not "muscular action," though I believe it is not always so, but I do say that you have as yet given not a particle of proof that it is so, while scattered through your paper is plenty of evidence which points to its being something quite different. Such are the cases when people hold the rod for the first time and have never seen a dowser work, yet the rod turns, over water, to their great astonishment, etc. etc.

Your conclusion that it is "clairvoyance" is a good provisional conclusion, but till we know what clairvoyance really is it explains nothing, and is merely another way of stating the *fact*.

I believe all true clairvoyance to be spirit impression, and that all true dowsing is the same—that is, when in either case it cannot be thought-transference, but even this I believe to be also, for the most part, if not wholly, spirit impression.—Believe me yours very truly,

ALFRED R. WALLACE.

### TO PROF. BARRETT

*Parkstone, Dorset. February 17, 1901.*

My dear Barrett,—I am rather sorry you wrote to any one of the Society for Psychical Research people about my being asked to be President, because I should certainly feel compelled to decline it. I never go, willingly, to London now, and should never attend meetings, so pray say no more about it. Besides, I am so widely known as a "crank" and a "faddist" that my being President would injure the Society, as much as Lord Rayleigh would benefit it, so pray do not put any obstacle in *his* way, though of course there is no necessity to beg him as a favour to be the successor of Sidgwick, Crookes and Myers....

### TO REV. J.B. HENDERSON

*Parkstone, Dorset. August 10, 1893.*

Dear Sir,—Although I look upon Christianity as originating in an unusual spiritual influx, I am not disposed to consider [it] as *essentially* different from those which originated other great religious and philanthropic movements. It is probable that in *your* sense of the word I am not a Christian.—Believe me yours very truly,

ALFRED R. WALLACE.

### TO MR. J.W. MARSHALL

*Parkstone, Dorset. March 6, 1894.*

My dear Marshall,—We were very much grieved to hear of your sad loss in a letter from Violet. Pray accept our sincere sympathy for Mrs. Marshall and yourself.

Death makes us feel, in a way nothing else can do, the mystery of the universe. Last autumn I lost my sister, and she was the only relative I have been with at the last. For the moment it seems unnatural and incredible that the living self with its special idiosyncrasies you have known so long can have left the body, still more unnatural that it should (as so many now believe) have utterly ceased to exist and become nothingness!

With all my belief in, and knowledge of, Spiritualism, I have, however, occasional qualms of doubt, the remnants of my original deeply ingrained scepticism; but my reason goes to support the psychical and spiritualistic phenomena in telling me that there *must* be a hereafter for us all....—Believe me yours very sincerely,

ALFRED R. WALLACE.

## TO DR. EDWIN SMITH

*Parkstone, Dorset. October 19, 1899.*

Dear Sir,—I know nothing of London mediums now. Nine-tenths of the alleged frauds in mediums arise from the ignorance of the sitters. The only way to gain any real knowledge of spiritualistic phenomena is to follow the course pursued in all science—study the elements before going to the higher branches. To expect proof of materialisation before being satisfied of the reality of such simpler phenomena as raps, movements of various objects, etc. etc., is as if a person began chemistry by trying to analyse the more complex vegetable products before he knew the composition of water and the simplest salts.

If you want to *know* anything about Spiritualism you should experiment yourself with a select party of earnest inquirers—personal friends. When you have thus satisfied yourself of the existence of a considerable range of the physical phenomena and of many of the obscurities and difficulties of the inquiry, you may use the services of public mediums, without the certainty of imputing every little apparent suspicious circumstance to trickery, since you will have seen similar suspicious facts in your private circle where you *knew* there was no trickery. You will find rules for forming private circles in some issues of *Light*. You can get them from the office of *Light*.—Yours very truly,

ALFRED R. WALLACE.

## PROF. BARRETT TO A.R. WALLACE

*6 De Vesci Terrace, Kingstown, Co. Dublin. November 3, 1905.*

My dear Wallace,— ... Just now I am engaged in a correspondence with the Secretaries of the Society for Psychical Research on the question of the Presidency for next year. I maintain that as a matter of duty to the Society you should be asked to accept the Presidency, though of course it would be impossible for you to be much more than an Honorary President, as we could not expect you often to come to London. I am anxious that in our records for future reference your Presidency should appear.... Podmore, who is proposed as President, represents the attitude of resolute incredulity, and I consider this line of action has been to some extent injurious to the S.P.R. Crookes supported my proposal, and so did Lodge, and so would Myers if he had lived. All this is of course between ourselves....

I have a vast amount of material unpublished on "dowsing" and am convinced the explanation is subconscious clairvoyance....—Yours very sincerely,

W.F. BARRETT.

## TO MRS. FISHER

*Broadstone, Wimborne. April 20, 1906.*

My dear Mrs. Fisher,—If you mean "honest" by "thoroughly reliable," there are plenty of such mediums, but if you mean those who give equally good results always, and to all persons, I should say there are none....

I am reading Herbert Spencer's "Autobiography" (just finished Vol. I.). I find it very interesting, though tedious in parts. I am glad I did not read it before I wrote mine. He certainly brings out his own character most strikingly, and a wonderful character it was. How extraordinarily little he owed either to teaching or to reading! I think he is best described as a "reasoning genius."— Yours very truly,

ALFRED R. WALLACE.

## LORD AVEBURY TO A.R. WALLACE

*48 Grosvenor Street, W. May 1, 1910.*

My dear Wallace,—I have been reading your biography with great interest. It must be a source of very pleasant memories to you to look back and feel how much you have accomplished.

It surprises me, however, how much we differ, and it is another illustration of the problems [?] of our (or rather I should say of my) intellect.

In some cases, indeed, the difference is as to facts.

You would, I am sure, for instance, find that you have been misinformed as to "thousands of dogs" being vivisected annually (p. 392).... As to Spiritualism, my difficulty is that nothing comes of it. What has been gained by your séances, compared to your studies?

I see you have a kindly reference to our parties at High Elms in old days, on which I often look back with much pleasure, but much regret also.

If you would give us the pleasure of another visit, *do* propose yourself, and you will have a very hearty welcome from yours very sincerely,

AVEBURY.

A lecture delivered by Prof. Barrett before the Quest Society in London, entitled "Creative Thought," was published by request, and as it discussed the subject of evolution and the impossibility of explaining the phenomena of life without a supreme Directing and Formative Force behind all the manifestations of life, he was anxious to have Wallace's criticisms. At that time he had not read Wallace's recently published work on a similar subject, and he was greatly surprised to find how closely his views agreed with those of the great naturalist.

## TO PROF. BARRETT

*Old Orchard, Broadstone, Wimborne. February 15, 1911.*

My dear Barrett,—Thanks for your proofs, which I return. It is really curious how closely your views coincide with mine, and how admirably and clearly you have expressed them. If it were not for your adopting throughout, as an actual fact, the (to me) erroneous theory of the "subconscious self," I should agree with every word of it. I have put "?" where this is prominently put forward, merely to let you know how I totally dissent from it. To me it is pure assumption, and, besides, proves nothing. Thanks for the flattering "Postscript," which I return with a slight suggested alteration.

Reviews have been generally very fair, complimentary and flattering. But to me it is very curious that even the religious reviewers seem horrified and pained at the idea that the Infinite Being does not actually do every detail himself, apparently leaving his angels, and archangels, his seraphs and his messengers, which seem to exist in myriads according to the Bible, to have no function whatever!—Yours very truly,

ALFRED R. WALLACE.

## PROF. BARRETT TO A.R. WALLACE

*6 De Vesci Terrace, Kingstown, Co. Dublin. February 18, 1911.*

My dear Wallace,— ... Thank you very much for your kind letter and comments. I have modified somewhat the phraseology as regards the "subliminal self." I think we really agree but use different terms. There *is* a hidden directive power, which works in conjunction with, and is temporarily part of, our own conscious self; but it is below the threshold of consciousness, or is a subliminal part of our self.

I should like to have come over to Broadstone expressly to ask your views on the parts you queried. For I have an immense faith in the soundness of your judgment, and in the accuracy of your views *in the long run*.

I should like also immensely to see you again and in your lovely home....— Yours ever sincerely,

W.F. BARRETT.

## TO PROF. BARRETT

*Old Orchard, Broadstone, Wimborne. February 20, 1911.*

My dear Barrett,—I wrote you yesterday on quite another matter, but having yours this morning in reply to my criticisms of your Address, I send a few lines of explanation. Most of my queries to your statements apply solely to your expressing them so positively, as if they were absolute certainties which no psychical researcher doubted. My main objection to the term "subliminal self" and its various synonyms is, that it is so dreadfully vague, and is an excuse for the assumption that a whole series of the most mysterious of psychical phenomena are held to be actually explained by it. Thus it is applied to explain all cases of apparent "possession," when the alleged "secondary self" has a totally different character, and uses the dialect of another social grade, from the normal self, sometimes even possesses knowledge that the real self could not have acquired, speaks a language that the normal self never learnt. All this is, to me, the most gross travesty of science, and I therefore object totally to the use of the term which is so vaguely and absurdly used, and of which no clear and rational explanation has ever been given.

You are now one of my oldest friends, and one with whom I most sympathise; and I only regret that we have seen so little of each other.—Yours very faithfully,

ALFRED R. WALLACE.

### TO MR. E. SMEDLEY

*Old Orchard, Broadstone, Dorset. October 2, 1911.*

Dear Mr. Smedley,—I am quite astonished at your wasting your money on an advertising astrologer. In the horoscope sent you there is not a single definite fact that would apply to you any more than to thousands of other men. All is vague, what "might be," etc. etc. It is just calculated to lead you on to send more money, and get in reply more words and nothing else....—Yours very truly,

ALFRED R. WALLACE.

A.R. WALLACE ADMIRING *EREMUS ROBUSTUS* about 1905.

# PART VII

### Characteristics

"There is a point of view so lofty or so peculiar that from it we are able to discern in men and women something more than and apart from creed and profession and formulated principle; which indeed directs and colours this creed and principle as decisively as it is in its turn acted on by them, and this is their character or humanity."—LORD MORLEY.

"As sets the sun in fine autumnal calm

So dost thou leave us. Thou not least but last

Link with that rare and gallant little band

Of seekers after truth, whose days, though past,

Shed lustre on the hist'ry of their land.

And thine, O Wallace, thine the added charm

Of modesty, thy mem'ry to embalm."—*Anonymous.*

(*Received with a bunch of lilies-of-the-valley, a few days after Dr. Wallace's death.*)

Addison somewhere says that modesty sets off every talent which a man can be possessed of. This was manifestly true of Alfred Russel Wallace. When, for instance, honours were bestowed upon him, he accepted or rejected them with the same good-humour and unspoilable modesty. To Prof. E.B. Poulton, whose invitation for the forthcoming Encæmia had been conveyed in Prof. Bartholomew Price's letter, he wrote:

*Godalming. May 28, 1889.*

My dear Mr. Poulton,—I have just received from Prof. B. Price the totally unexpected offer of the honorary degree of D.C.L. at the coming Commemoration, and you will probably be surprised and *disgusted* to hear that I have declined it. I have to thank you for your kind offer of hospitality during the ceremony, but the fact is, I have at all times a profound distaste of all public ceremonials, and at this particular time that distaste is stronger than ever. I have never recovered from the severe illness I had a year and a half ago, and it is in hopes of restoring my health that I have let my cottage here and have taken another at Parkstone, Dorset, into which I have arranged to move on Midsummer Day. To add to my difficulties, I have work at examination papers for the next two or three weeks, and also a meeting (annual) of our Land Nationalisation Society, so that the work of packing my books and other things and looking after the plants which I have to move from my garden will have to be done in a very short time. Under these circumstances it would be almost impossible for me to rush away to Oxford except under absolute compulsion, and to do so would be to render a ceremony which at any time would be a trial, a positive punishment.

Really the greatest kindness my friends can do me is to leave me in peaceful obscurity, for I have lived so secluded a life that I am more and more disinclined to crowds of any kind. I had to submit to it in America, but then I felt exceptionally well, whereas now I am altogether weak and seedy and not at all up to fatigue or excitement.—Yours very faithfully,

ALFRED R. WALLACE.

Prof. Poulton pressed him to reconsider his decision, and he reluctantly gave way.

*Godalming. June 2, 1889.*

My dear Mr. Poulton,—I am exceedingly obliged by your kind letters, and I will say at once that if the Council of the University should again ask me to accept the degree, to be conferred in the autumn, as you propose, I could not possibly refuse it. At the same time I hope you will not in any way urge it upon them, as I really feel myself too much of an amateur in Natural History and altogether too ignorant (I left school—a bad one—finally, at fourteen) to receive honours from a great University. But I will say no more about that.—Yours very faithfully,

A.R. WALLACE.

In due course he received the degree. "On that occasion," says Professor Poulton, "Wallace stayed with us, and I was anxious to show him something of Oxford; but, with all that there is to be seen, one subject alone absorbed the whole of his interest—he was intensely anxious to find the rooms where Grant Allen had lived. He had received from Grant Allen's father a manuscript poem giving a picture of the ancient city dimly seen by midnight from an undergraduate's rooms. With the help of Grant Allen's college friends we were able to visit every house in which he had lived, but were forced to conclude that the poem was written in the rooms of a friend or from an imaginary point of view."

His friend Sir W.T. Thiselton-Dyer, with others, was promoting his election to the Royal Society, and wrote to him:

### SIR W.T. THISELTON-DYER TO A.R. WALLACE

*Kew. October 23, 1892.*

Dear Mr. Wallace,— ... When you were at Kew this summer I took the liberty of saying that it would give great pleasure to the Fellows of the Royal Society if you would be willing to join their body. I understood you to say that it would be agreeable to you. I now propose to comply with the necessary formalities. But before doing so it will be proper to ask for your formal consent. You will then, as a matter of course, be included in the next annual election.

Will you forgive me if I am committing any indiscretion in saying that I have good authority for adding (though I suppose it can hardly be stated officially at this stage) that no demand will ever be made upon you for a subscription?—Believe me yours sincerely,

W.T. THISELTON-DYER.

### SIR W.T. THISELTON-DYER TO A.R. WALLACE

*Kew. January 12, 1893.*

Dear Mr. Wallace,— ... I was very vexed to hear that I had misunderstood your wishes about the Royal Society. Of course, the matter must often have presented itself to your mind, and I confess that it argued a little presumption on the part of a person like myself, so far inferior to you in age and standing, to think that you would yield to my solicitation.

I was obliged for my health to go to Eastbourne, and there I had the pleasure of seeing Mr. Huxley, who, you will be glad to hear, is wonderfully well, and an ardent gardener! His present ambition is to grow every possible saxifrage.

I told him that I had had the audacity to approach you on the subject of the Royal Society. He heartily approved, and expressed the strongest opinion that unless you had some insuperable objection you ought to yield. All of us who belong to the R.S. have but one wish, which is that it should stand before the public as containing all that is best and worthiest in British Science. As long as men like you stand aloof, that cannot be said. Lately we have been exposed to some very ill-natured attacks: we have been told that we are professional, and not discoverers. Well, this is all the more reason for your not holding aloof from us. I wish you would think it over again. Huxley went the length of saying that to him it seemed a plain duty. But this is language I do not like to use.

As to attending the meetings or taking part in the work of the Society, that is immaterial. Darwin never did either, though he did once come to one of the evening receptions, and enjoyed it immensely.

In writing as I do I am not merely expressing my own opinions, but those of many others of my own standing who are keenly interested in the matter.

It is not a great matter to ask. I have the certificate ready. You have but to say the word. You will be put to no trouble or pecuniary responsibility. That my father-in-law arranged, long ago.

To dissociate yourself from the R.S. really amounts nowadays to doing it an injury. And I am sure you do not wish that.

With all good wishes, believe me yours sincerely,

W.T. THISELTON-DYER.

## TO SIR W.T. THISELTON-DYER

*Parkstone, Dorset. January 17, 1893.*

Dear Mr. Thiselton-Dyer,—I have been rather unwell myself the last few days or should have answered your very kind letter sooner. I feel really overpowered. I cannot understand why you or anyone should care about my being an F.R.S., because I have really done so little of what is usually considered scientific work to deserve it. I have for many years felt almost ashamed of the amount of reputation and honour that has been awarded me. I can understand the general public thinking too highly of me, because I know that I have the power of clear exposition, and, I think, also, of logical reasoning. But all the work I have done is more or less amateurish and founded almost wholly on other men's observations; and I always feel myself dreadfully inferior to men like Sir J. Hooker, Huxley, Flower, and scores of younger men who have extensive knowledge of whole departments of biology of which I am totally ignorant. I do not wish, however, to be thought ungrateful for the many honours that have been given me by the Royal and other Societies, and will therefore place myself entirely in your hands as regards my election to the F.R.S.

I am much pleased to hear that Huxley has taken to gardening. I have no doubt he will do some good work with his saxifrages. For myself the personal attention to my plants occupies all my spare time, and I derive constant enjoyment from the mere contemplation of the infinite variety of forms of leaf and flower, and modes of growth, and strange peculiarities of structure which are the source of fresh puzzles and fresh delights year by year. With best wishes and many thanks for the trouble you are taking on my behalf, believe me yours very faithfully,

ALFRED R. WALLACE.

In 1902 the *Standard* announced that the degree of D.C.L. was to be conferred upon him by the University of Wales. He wrote to Miss Dora Best, who had sent him the information:

I have not seen the *Standard*. But I suppose it is about the offer of a degree by the University of Wales. You will not be surprised to hear that I have declined it "with thanks." The bother, the ceremony, the having perhaps to get a blue or yellow or scarlet gown! and at all events new black clothes and a new topper! such as I have not worn this twenty years. Luckily I had a good excuse in having committed the same offence before. Some ten years back I declined the offer of a degree from Cambridge, so that settled it.

P.S.—Having already degrees two—LL.D. (Dublin) and D.C.L. (Oxford)—I might have quoted Shakespeare: "To gild refined gold, to paint the lily," etc. But I didn't!—A.R.W.

In 1908 he received the Order of Merit, the highest honour conferred upon him. To his friend Mrs. Fisher he wrote:

Dear Mrs. Fisher,—Is it not awful—two more now! I should think very few men have had three such honours within six months! I have never felt myself worthy of the Copley Medal—and as to the Order of Merit—to be given to a red-hot Radical, Land Nationaliser, Socialist, Anti-Militarist, etc. etc. etc., is quite astounding and unintelligible!...

There is another thing you have not heard yet, but it will be announced soon. Sir W. Crookes, as Secretary of the Royal Institution, wrote to me two weeks back asking me very strongly to give them a lecture at their opening meeting (third week in January) appropriate to the Jubilee of the "Origin of Species." I was very unwell at the time—could eat nothing, etc.—and was going to decline positively, having nothing more to say! But while lying down, vaguely thinking about it, an idea flashed upon me of a new treatment of the whole subject of Darwinism, just suitable for a lecture to a R.I. audience. I felt at once there was something that ought to be said, and that I should like to say—so I actually wrote and accepted, provisionally. My voice has so broken that unless I can improve it I fear not being heard, but Crookes promised to read it either wholly, or leaving to me the opening and concluding paragraphs. I was very weak—almost a skeleton—but I am now getting much better. But finishing up the "Spruce" book, and now all these honours and congratulations and letters, etc., are giving me much work, yet I am getting strong again, and really hope to do this "lecture" as my last stroke for Darwinism against the Mutationists and Mendelians, but much more effective, I hope, than my article in the August *Contemporary Review*, though that was pretty strong.—Yours very sincerely,

ALFRED R. WALLACE.

How more than true "Sunlight's"[65] words have come, "You will come out of the hole! You will be more in the world. You will have satisfaction, retrospection, and work"! Literally fulfilled!—A.R.W.

### And to Mr. F. Birch:

*December 30, 1908.*

Dear Fred,— ... I received a letter from Lord Knollys—the King's Private Secretary—informing me that His Majesty proposed to offer me the Order of Merit, among the Birthday honours! This is an "Order" established by the present King about eight years ago, solely for "merit"—whether civil or military—it is a pity it was not civil only, as the military have so many distinctions already. So I had to compose a very polite letter of acceptance and thanks, and then later I had to beg to be excused (on the ground of age and delicate health) from attending the investiture at Buckingham Palace (on December 14th), when Court dress—a kind of very costly livery—is obligatory! and I was kept for weeks waiting. But at last one of the King's Equerries, Col. Legge (an Earl's son), came down here about two weeks ago bringing the Order, which is a very handsome cross in red and blue enamel and gold—rich colours—with a crown above, and a rich ribbed-silk blue and crimson riband to hang it round the neck! Col. Legge was very pleasant, stayed half an hour, had some tea, and showed us how to wear it. So I shall be in duty bound to wear it on the only public occasion I shall be seen again (in all probability), when I give (or attempt to give) my lecture.[66] Then, I had a letter from Windsor telling me that chalk portraits of all the members of the Order were to be taken for the collections in the Library, and a Mr. Strang came and stayed the night, and in four hours completed a very good life-size head, in coloured chalk, and so far, so good!—Yours very sincerely,

ALFRED R. WALLACE.

Wallace regarded "Sunlight's" prophecy about "retrospection" as being fulfilled in 1904, when he received the invitation of Messrs. Chapman and Hall to begin collecting material for his autobiography which was subsequently published in two large volumes, under the title of "My Life."

Referring to this work he wrote to Mrs. Fisher:

*Broadstone, Dorset. April* 17, 1904.

Dear Mrs. Fisher,—Thanks for your remarks on what an autobiography ought to be. But I am afraid I shall fall dreadfully short. I seem to remember nothing but ordinary facts and incidents of no interest to anyone but my own family. I do not feel myself that anything has much influenced my character or abilities, such as they are. Lots of things have given me opportunities, and those I can state. Also other things have directed me into certain lines, but I can't dilate on these; and really, with the exception of Darwin and Sir Charles Lyell, I have come into close relations with hardly any eminent men. All my doings and surroundings have been commonplace!

I am now just reading a charming and ideal bit of autobiography—Robert Dale Owen's "Threading my Way." If you have not read it, do get it (published by Trübner and Co. in 1874). It is delightful. So simple and natural throughout. But his father was one of the most wonderful men of the nineteenth century—Robert Owen of New Lanark—and this book gives the true history of his great success. Then R.D. Owen met Clarkson and heard from his own lips how he worked to abolish the slave trade.

Then he had part of his education at Hofwyl under Fellenberg, an experiment in education and self-government wonderfully original and successful. He afterwards worked at "New Harmony" with his father, and met during his life almost all the most remarkable people in England and America.

This book only contains the first twenty-seven years of his life and I am afraid he never completed it. Such a book makes me despair!—Yours very sincerely,

ALFRED R. WALLACE.

When "My Life" was published, he wrote to the same old and valued friend:

*Broadstone, Wimborne. November 7, 1905.*

My dear Mrs. Fisher,—The reviewers are generally very fair about the fads except a few. The *Review* invents a new word for me—I am an "anti-body"; but the *Outlook* is the richest: I am the one man who believes in Spiritualism, phrenology, anti-vaccination, and the centrality of the earth in the universe, whose life is worth writing. Then it points out a few things I am capable of believing, but which everybody else knows to be fallacies, and compares me to Sir I. Newton writing on the prophets! Yet of course he praises my biology up to the skies—there I am wise—everywhere else I am a kind of weak, babyish idiot! It is really delightful!

Only one is absolutely savage about it all—the *Liverpool Daily Post and Mercury*. The reviewer devotes over three columns almost wholly to the fads—as to all of which he evidently knows absolutely nothing, but he is cocksure that I am always wrong!....—Yours very sincerely,

ALFRED R. WALLACE.

He always thought that he was deficient in the gift of humour: "I am," he wrote to Mr. J.W. Marshall (May 6, 1905), "still grinding away at my autobiography. Have got to my American lecture tour, and hope to finish by about Sept. but have such lots of interruptions. I am just reading Huxley's Life. Some of his letters are inimitable, but the whole is rather monotonous. I find there is a good deal of variety in my life if I had but the gift of humour! Alas! I could not make a joke to save my life. But I find it very interesting." "Unless somebody," he wrote to Miss Evans, "can make me laugh just before the critical moment I always have a horrid expression in photographs." Yet another observant friend remarked that "he had a keen sense of humour. It was always his boyish joyous exuberance which touched me. He never grew old. When I had sat with him an hour he was a young man, he became transfigured to me." ... "The last time I saw Dr. Wallace," writes Prof. T.D.A. Cockerell of Colorado, "was immediately after the Darwin Celebration at Cambridge in 1909. I was the first to give him the details concerning it, and vividly remember how interested he was, and how heartily he laughed over some of the funny incidents, which may not as yet be told in print. One of his most prominent characteristics was his keen sense of humour, and his enjoyment of a good story." In the summer of 1885 he spent a holiday with Prof. Meldola at Lyme Regis. "After our ramble," said the Professor, "we used to spend the evenings indoors, I reading aloud the 'Ingoldsby Legends,' which Wallace richly enjoyed. His humour was a delightful characteristic. 'The inimitable puns of T. Hood were,' he said, 'the delight of my youth, as is the more recondite and fantastic humour of Mark Twain and Lewis Carroll in my old age.'"

Wallace loved to give time and trouble in aiding young men to start in life, especially if they were endeavouring to become naturalists. He sent them letters of advice, helped them in the choice of the right country to visit, and gave them minute practical instructions how to live healthily and to maintain themselves. He put their needs before other and more fortunate scientific workers and besought assistance for them.

"The central secret of his personal magnetism lay in his wide and unselfish sympathy," writes Prof. Poulton.[67] "It might be thought by those who did not know Wallace that the noble generosity which will always stand as an example before the world was something special—called forth by the illustrious man with whom he was brought in contact. This would be a great mistake. Wallace's attitude was characteristic, and characteristic to the end of his life.

"A keen young naturalist in the North of England, taking part in an excursion to the New Forest, called on Wallace and confided to him the dream of his life—a first-hand knowledge of tropical nature. When I visited 'Old Orchard' in the summer of 1903, I found that Wallace was intently interested in two things: his garden, and the means by which his young friend's dream might best be realised. The subject was referred to in seventeen letters to me; it formed the sole topic of some of them. It was a grand and inspiring thing to see this great man identifying himself heart and soul with the interests of one—till then a stranger—in whom he recognised the passionate longings of his own youth. By the force of sympathy he re-lived in the life of another the splendid years of early manhood."

The late Prof. Knight recalled meeting him at the British Association in Dundee, during the year 1867, when Wallace was his guest for the usual time of the gathering. He wrote:

I, and everyone else who then met him at my house, were struck, as no one could fail to be, by his rare urbanity, his social charm, his modesty, his unobtrusive strength, his courtesy in explaining matters with which he was himself familiar but those he conversed with were not; and his abounding interest, not only in almost every branch of Science, but in human knowledge in all its phases, especially new ones. He was a many-sided scientific man, and had a vivid sense of humour. He greatly enjoyed anecdote, as illustrative of character. During those days he talked much on the fundamental relations between Science and Philosophy, as well as on the connection of Poetry with both of them. When he left Dundee he went to Kenmore, that he might ascend Ben Lawers in search of some rare ferns.

In 1872 I saw him, after meeting Thomas Carlyle and Dean Stanley at Linlathen, when Darwin's theory was much discussed, and when our genial host—Mr. Erskine—talked so dispassionately but decidedly against evolution as explanatory of the rise of what was new. A little later in the same year Matthew Arnold discussed the same subject with some friends at the Athenæum Club, defending the chief aim of Darwin's theory, and enlarging from a different point of view what Wallace had done in the same direction. I remember well that he characterised the two men as fellow-workers, not as followers, or in any sense as copyists. Wallace's versatility not only continued, but grew in many ways with the advance of years. It was seen in his appreciation of the value of historical study. Quite late in life he wrote: "The nineteenth century is quite as wonderful in the domain of History as in that of Science." Comparatively few know, or remember, that he and his young brother Herbert—on whom he left an interesting chapter *in memoriam*—both wrote verses, some of which were of real value.

It may be safely said that few scientific men have sympathetically entered into bordering territories and therein excelled. The whole field of psychical research was familiar to him, and he might have been a leader in it.

My last meeting with him was at his final home, the "Old Orchard," Broadstone, in 1909. I was staying at Boscombe in Hants, and he asked me to "come and see his garden, while we talked of past days." He had then the freshness of boyhood, blent with the mellow wisdom of age.—W.A.K.

The eminent naturalist and traveller, Dr. Henry O. Forbes, who later explored the greater part of the lands visited by Wallace, contributes the following appreciation of the latter's scientific work:

As a traveller, explorer and working naturalist, Wallace will always stand in the first rank, compared even with the most modern explorers. It ought not to be forgotten, however, how great were the difficulties, the dangers and the cost of travel fifty years ago, compared with the facilities now enjoyed by his successors, who can command steam and motor transport to wellnigh any spot on the coasts of the globe, and who have to their hand concentrated and preserved foods, a surer knowledge of the causes of tropical diseases, and outfits of non-perishable medicines sufficient for many years within the space of a few cubic inches. Commissariat and health are the keys to all exploration in uncivilised regions. Wallace accomplished his work on the shortest of commons and lay weeks at a time sick through inability to replenish his medical stores.

He was no mere "trudger" over new lands. Where those before him, and even many after him, have been able to see only sterile objects, his discerning eyes perceived everywhere a meaning in the varying modes of organic life, and in response to his sympathetic mind Nature revealed to him more of her multitudinous secrets than to most others. Wallace's Amazonian travels were far from unfruitful, in spite of the irreparable loss he sustained in the burning of his notes and the bulk of his collections in the vessel by which he was returning home; but it was in the Malay Archipelago that his most celebrated years of investigation were passed, which marked him as one of the greatest naturalists of our time. As a methodical natural history collector—which is "the best sport in the world" according to Darwin—he has never been surpassed; and few naturalists, if any, have ever brought together more enormous collections than he. The mere statement, taken from his "Malay Archipelago," of the number of his captures in the Archipelago in six years of actual collecting, exceeding 125,000 specimens—a number greater than the entire contents of many large museums—still causes amazement. The value of a collection, however, depends on the full and accurate information attached to each specimen, and from this point of view only a few collections, including Darwin's and Bates's, have possessed the great scientific value of his.

Wallace's Eastern explorations included nearly all the large and the majority of the smaller islands of the Archipelago. Many of them he was the first naturalist to visit, or to reside on. Ceram, Batjian, Buru, Lombok, Timor, Aru, Ke and New Guinea had never been previously scientifically investigated. When in 1858 "the first and greatest of the naturalists," as Dr. Wollaston styles Wallace, visited New Guinea, it was "the first time that any European had ventured to reside alone and practically unprotected on the mainland of this country," which, dangerous as it is now in the same regions, was infinitely more so then. Of the journals of his voyagings, "The Malay Archipelago" will always be ranked among the greatest narratives of travel. The fact that this volume has gone through a dozen editions is witness to its extraordinary popularity among intelligent minds, and hardly supports the belief that his scientific work has been forgotten. Nor can this popularity be a matter of much surprise, for few travellers have possessed Wallace's powers of exposition, his lucidity and charm of style. Professor Strasburger of Bonn has declared that through "The Malay Archipelago" "a new world of scientific knowledge" was unfolded before him. "I feel it ... my duty," he adds, "to proclaim it with gratitude." Wallace's narrative has attracted during the past half-century numerous naturalists to follow in his tracks, many of whom have reaped rich aftermaths of his harvest; but certain it is that no explorer in the same, if in any other, region has approached his eminence, or attained the success he achieved.

As a systematic zoologist, Wallace took no inconsiderable place; his *métier*, however, was different. He described, nevertheless, large sections of his Lepidoptera and of his birds, on which many valuable papers are printed in the *Transactions* of the learned societies and in various scientific periodicals. Of the former, special mention may be made of that on variation in the "Papilionidæ of the Malayan Region," of which Darwin has recorded: "I have never in my life been more struck by any paper." Of the latter, reference may be drawn to his account of the "Pigeons of the Malay Archipelago" and his paper on the "Passerine Birds," in which he proposed an important new arrangement of the families of that group (used later in his "Geographical Distribution") based on the feathering of their wings. Without a lengthy search through the zoological records, it would be impossible to say how many species Wallace added to science; but the constant recurrence in the Catalogue of Birds in the British Museum of "wallacei" as the name bestowed on various new species by other systematists, and of "Wallace" succeeding those scientifically named by himself, is an excellent gauge of their very large number.

In the field of anthropology Wallace could never be an uninterested spectator. He took a deep interest, he tells us, in the study of the various races of mankind. His accounts of the Amazonian tribes suffered greatly by the loss of his journals; but of the peoples of the Malay Archipelago he has given us a most interesting narrative, detailing their bodily and mental characteristics, and showing how their distribution accorded with that of the fauna on the opposite sides—Malays to the West, Papuans to the East—of Wallace's Line. If fuller investigation of the New Guinea tribes requires some modification in regard to their origin, his observations, as broadly outlined then, remain true still. His opinions on the origin of the Australian aborigines—that they were a low and primitive type of Caucasian race—which, when first promulgated, were somewhat sceptically received, are now those accepted by many very competent anthropologists.

Wallace's contributions to Geographical Science were only second in importance to those he so pre-eminently made to biology. Though skilled in the use of surveying instruments, he did little or no map-making—at all times a laborious and lengthy task—for, with more important purposes in his mind, he could not spare the time, nor did the limitations to his movements permit any useful attempt. Yet he did pure geographical work quite as important. The value of the comparative study of the flora and fauna of neighbouring regions, the great differences in the midst of much likeness between the organic life of neighbouring land masses, was a subject that was always in Wallace's mind during his exploration of the Amazon Valley, for he perceived that the physical geography and the distribution of these animals and plants were of the greatest service in elucidating their history where the geological record was defective. As is well known, the visual inspection of the geological structure of tropical countries is always difficult and often impossible to make out because of the dense vegetation upon the surface and even the faces of the river gorges. But for the loss of his collections and notes we should have had from Wallace's pen a Physical History of the Amazon. This loss was, however, amply made up by his very original contributions to the geography of the Malay Archipelago. "The Zoological Geography of the Malay Archipelago" and "The Physical Geography of the Malay Archipelago" (written on Eastern soil, with the texts of his discourses around him) were the forerunners of his monumental "Geographical Distribution of Animals," elaborated in England after his return. "To the publication of the 'Geographical Distribution of Animals' we owe the first scientific study of the distribution of organic life on the globe, which has broadened ever since, and continues to interest students daily; his brilliant work in Natural History and Geography ... is universally honoured," are the opinions of Dr. Scott speaking as President of the Linnean Society of London.

One of Wallace's most important contributions to the physical geography of the Malay region was his discovery of the physical differences between the western and the eastern portions of the Archipelago; i.e. that the islands lying to the east of a line running north from the middle of the Straits of Bali and outside Celebes were fragments of an ancient and larger Australian continent, while those to the western side were fragments of an Asiatic continent. This he elucidated by recognising that the flora and fauna on the two sides of the line, close though these islands approached each other, were absolutely different and had remained for ages uncommingled. This line was denominated "Wallace's Line" by Huxley, and this discovery alone would have been sufficient to associate his name inseparably with this region of the globe.—H.O.F.

Like Darwin, Wallace gave excessive attention to the suggestions and criticisms of people who were obviously ignorant of the subjects about which they wrote. He was never impatient with honest ignorance or considered the lowly position of his correspondents. He replied to all letters of inquiry (and he received many from working men), and always gave his best knowledge and advice to anyone who desired it. There was not the faintest suggestion of the despicable sense of superiority about him.

"I had, of course, revelled in 'The Malay Archipelago' when a boy," says Prof. Cockerell, "but my first personal relations with Dr. Wallace arose from a letter I wrote him after reading his 'Darwinism,' then (early in 1890) recently published. The book delighted me, but I found a number of little matters to criticise and discuss, and with the impetuosity of youth proceeded to write to the author, and also to send a letter on some of the points to *Nature*. I have possibly not yet reached years of discretion, but in the perspective of time I can see with confusion that what I regarded as worthy zeal might well have been characterised by others as confounded impudence. In the face of this, the tolerance and kindness of Dr. Wallace's reply is wholly characteristic: 'I am very much obliged to you for your letter containing so many valuable emendations and suggestions on my "Darwinism." They will be very useful to me in preparing another edition. Living in the country with but few books, I have often been unable to obtain the *latest* information, but for the purpose of the argument the facts of a few years back are often as good as those of to-day—which in their turn will be modified a few years hence.... You appear to have so much knowledge of details in so many branches of natural history, and also to have thought so much on many of the more recondite problems, that I shall be much pleased to receive any further remarks or corrections on any other portions of my book.' This letter, written to a very young and quite unknown man in the wilds of Colorado, who had merely communicated a list of more or less trifling criticisms, can only be explained as an instance of Dr. Wallace's eagerness to help and encourage beginners. It did not occur to him to question the propriety of the criticisms, he did not write as a superior to an inferior; he only saw what seemed to him a spark of biological enthusiasm, which should by all means be kindled into flame. Many years later, when I was at his house, he produced with the greatest delight some letters from a young man who had gone to South America and was getting his first glimpse of the tropical forest. What discoveries he might make! What joy he must have on seeing the things described in the letter, such things as Dr. Wallace himself had seen in Brazil so long ago!"

Wallace's critical faculty was always keen and vigilant. Unlike some critics, however, he relished genuine and well-informed criticism of his own writings. Flattery he despised; whilst the charge of dishonesty aroused strongest resentment. Deceived he might be, but he required clear proof that his own eyes and ears had led him astray. Romanes, who had propounded the forgotten theory of physiological selection, charged Wallace with adopting it as his own. This was not only untrue, it was ridiculous; and Wallace, after telling him so and receiving no apology, dropped him out of his recognition. During Romanes' illness Mr. Thiselton-Dyer wrote to Wallace and sought to bring about a reconciliation, and Wallace replied:

*Parkstone, Dorset. September 26, 1893.*

My dear Thiselton-Dyer,—I am sorry to hear of Romanes' illness, because I think he would have done much good work in carrying out experiments which require the leisure, means and knowledge which he possesses. I cannot, however, at all understand his wishing to have any communication from myself. I do not think I ever met Romanes in private more than once, when he called on me more than twenty years ago about some curious psychical phenomena occurring in his own family; and perhaps half a dozen letters—if so many—may have passed between us since. There is therefore no question of personal friendship disturbed. I consider, however, that he made a very gross misstatement and personal attack on me when he stated, both in English and American periodicals, that in my "Darwinism" I adopted his theory of "physiological selection" and claimed it as my own, and that my adoption of it was "unequivocal and complete." This accusation he supported by such a flood of words and quotations and explanations as to obscure all the chief issues and render it almost impossible for the ordinary reader to disentangle the facts. I told him then that unless he withdrew this accusation as publicly as he had made it I should decline all future correspondence with him, and should avoid referring to him in any of my writings.

This is, of course, very different from any criticism of my theories; that, or even ridicule, would never disturb me; but when a man has made an accusation of literary and scientific dishonesty, and has done all he can to spread this accusation over the whole civilised world, my only answer can be—after showing, as I have done (*see Nature*, vol. xliii., pp. 79 and 150), that his accusations are wholly untrue—to ignore his existence.

I cannot believe that he can want any sympathy from a man he says has wilfully and grossly plagiarised him, unless he feels that his accusations were unfounded. If he does so, and will write to me to that effect (for publication, if I wish, after his death), I will accept it as full reparation and write him such a letter as you suggest.—Believe me yours very faithfully,

ALFRED R. WALLACE.

# SIR W.T. THISELTON-DYER TO A.R. WALLACE

*Kew. September 27, 1897.*

Dear Mr. Wallace,—I am afraid I have been rather guilty of an impertinence which I hope you will forgive.

Romanes is an old acquaintance of mine of many years' standing. Personally, I like him very much; but for his writings I confess I have no great admiration.

Pray believe me I had no mission of any sort on his part to write to you. But I feel so sorry for him that when he told me how much he regretted that he did not stand well with you, I could not resist writing to tell you of the calamities that have befallen him.

I must confess I was in total ignorance of what you tell me. I don't see how, under the circumstances, you can do anything. I was never more surprised in my life, in fact, than when I read your letter. The whole thing is too childishly preposterous.

Romanes laments over *me* because he says I wilfully misunderstand his theory. The fact is, poor fellow, that I do not think he understands it himself. If his life had been destined to be prolonged I should have done all in my power to have induced him to occupy himself more with observation and less with mere logomachy.

I cannot get him to face the fact that natural hybrids are being found to be more and more common amongst plants. At the beginning of the century it was supposed that there were some sixty recognisable species of willows in the British Isles: now they are cut down to about sixteen, and all the rest are resolved into hybrids.—Ever sincerely,

W.T. THISELTON-DYER.

Wallace was a seeker after Truth who was never shy of his august mistress, whatever robes she wore. "I feel within me," wrote Darwin to Henslow, "an instinct for truth, or knowledge, or discovery, of something of the same nature as the instinct of virtue." This was equally true of Wallace. He had a fine reverence for truth, beauty and love, and he feared not to expose error. He paid no respect to time-honoured practices and opinions if he believed them to be false. Vaccination came under his searching criticism, and in the face of nearly the whole medical faculty he denounced it as quackery condemned by the very evidence used to defend it. He very carefully examined the claims of phrenology, which had been laughed out of court by scientific men, and he came to the conclusion that "in the present (twentieth) century phrenology will assuredly attain general acceptance. It will prove itself to be the true science of the mind. Its practical uses in education, in self-discipline, in the reformatory treatment of criminals, and in the remedial treatment of the insane, will gain it one of the highest places in the hierarchy of the sciences; and its persistent neglect and obloquy during the last sixty years of the nineteenth century will be referred to as an example of the almost incredible narrowness and prejudice which prevailed among men of science at the very time they were making such splendid advances in other fields of thought and discovery."[68]

Wallace was not even scared out of his wits by ghosts, for, unlike Coleridge, he believed in them although he thought he had seen many. Whether truth came from the scaffold or the throne, the séance or the sky, it did not alter the truth, and did not prejudice or overbear his judgment. He shed his early materialism (which temporarily took possession of him as it did of many others as a result of the shock following the overwhelming discoveries of that period) when he was brought face to face with the phenomena of the spiritual kingdom which withstood the searching test of his keen observation and reasoning powers. Prejudices, preconceived notions, respect for his scientific position or the opinions of his eminent friends or the reputation of the learned societies to which he belonged—all were quietly and firmly put aside when he saw what he recognised to be the truth. If his fellow-workers did not accept it, so much the worse for them. He stood four-square against the onslaught of quasi-scientific rationalism, which once threatened to obliterate all the ancient landmarks of morality and religion alike. He made mistakes, and he admitted and corrected them, because he verily loved Truth for her own sake. And to the very end of his long life he kept the windows of his soul wide open to what he believed to be the light of this and other worlds.

He was, then, a man of lofty ideals, and his idealism was at the base of his opposition to the materialism which boasted that Natural Selection explained all adaptation, and that Physics could give the solution of Huxley's poser to Spencer: "Given the molecular forces in a mutton chop, deduce Hamlet and Faust therefrom," and which regarded mind as a quality of matter as brightness is a quality of steel, and life as the result of the organisation of matter and not its cause.

"We have ourselves," wrote Prof. H.F. Osborn in an account of Wallace's scientific work which Wallace praised, "experienced a loss of confidence with advancing years, an increasing humility in the face of transformations which become more and more mysterious the more we study them, although we may not join with this master in his appeal to an organising and directing principle." But profound contemplation of nature and of the mind of man led Wallace to belief in God, to accept the Divine origin of life and consciousness, and to proclaim a hierarchy of spiritual beings presiding over nature and the affairs of nations. "Whatever," writes Dr. H.O. Forbes, "may be the last words on the deep and mysterious problems to which Wallace addressed himself in his later works, the unquestioned consensus of the highest scientific opinion throughout the world is that his work has been for more than half a century, and will continue to be, a living stimulus to interpretation and investigation, a fertilising and vivifying force in every sphere of thought."

It is perhaps unprofitable to go further than in previous chapters into his so-called heresies—political, scientific or religious. Yet we may imitate his boldness and ask whether he was not, perhaps, in advance of his age and whether his heresies were not shrewd anticipations of some truth at present but partially revealed. Take the example of Spiritualism, which, I suppose, has more opponents than anti-vaccination. No one can overlook the fact that Spiritualism has many scientific exponents—Myers, Crookes, Lodge, Barrett and others. Prejudices against Spiritualism are as unscientific as the credulity which swallows the mutterings of every medium. Podmore's two ponderous volumes on the History of Spritualism are marred by an obvious anxiety to make the very least, if not the very worst, of every phenomenon alleged to be spiritualistic. That kind of deliberate and obstinate blindness which prided itself on being the clear cold light of science Wallace scorned and denounced. He did not insist upon spiritualistic manifestations shaping themselves according to his own predesigned moulds in order to be investigated. He watched for facts whatever form they assumed. He fully recognised that the phenomena he saw and heard could be easily ridiculed, but behind them he as fully believed that he came into contact with spiritual realities which remain, and which led him to other explanations of the higher faculties of man and the origin of life and consciousness than were acceptable to the materialistic followers of Haeckel, Büchner and Huxley. And who dares dogmatically to assert in the name of science and in the second decade of the twentieth century, when the deeper meanings of evolution are being revealed, and the philosophy of Bergson is spoken about on the housetops, that he was wrong? In these views may he not become the peer of Darwin?

At first blush it may seem to be a bad example of special pleading to attempt to discover the reason for his opposition to vaccination in his idealism. But it is not far from the truth. He believed in a Ministry of Public Health, that doctors should be servants of the State, and that they should be paid according as they kept people well and not ill. Health is the natural condition of the human body when it is properly sustained and used. And chemicals, even in sickness, are of less importance than fresh air, light and proper food. He ridiculed, too, the notion of unhealthy places. "It is like," he wrote to Mr. Birch, "the old idea that every child must have measles, and the sooner the better." To the same correspondent, who was contemplating going into virgin forests and who expressed his fear of malaria, he replied: "There is no special danger of malaria or other diseases in a dense forest region. I am sure this is a delusion, and the dense virgin forests, even when swampy, are, in a state of nature, perfectly healthy to live in. It is man's tampering with them, and man's own bad habits of living, that render them unhealthy. Having now gone over all Spruce's journals and letters during his twelve years' life in and about the Amazonian forests, I am sure this is so. And even where a place is said to be notoriously 'malarious,' it is mostly due not to infection only but to predisposition due to malnutrition or some bad mode of living. A person living healthily may, for the most part, laugh at such terrors. Neither I nor Spruce ever got fevers when we lived in the forests and were able to get wholesome food." "Health," he said to the present writer, "is the best resistant to disease, and not the artificial giving of a mild form of a disease in order to render the body immune to it for a season. Vaccination is not only condemned upon the statistics which are used to uphold it, but it is a false principle—unscientific, and therefore doomed to fail in the end." Besides which, he believed in mental healing, and had recorded definite and certain benefit from spiritual "healers." And he reminded himself that amongst doctors (witness the blind opposition encountered by Lister's discoveries) were found from time to time not a few enemies of the true healing art, and obstinate defenders of many forms of quackery.

Wallace made no claim to be an original investigator. He knew his limitations, and said again and again that he could not have conducted the slow and minute researches or have accumulated the vast amount of detailed evidence to which Darwin, with infinite patience, devoted his life. He was genuinely glad that it had not fallen to his lot to write "The Origin of Species." He felt that his chief faculty was to reason from facts which others discovered. Yet he had that original insight and creative faculty which enabled him to see, often as by flashlight, the explanation which had remained hidden from the eyes of the man who was most familiar with the particular facts, and he elaborated it with quickening pulse, anxious to put down the whole conception which filled his mind lest some portion of it should escape him. Therein lay one secret of his great genius. He often said that he was an idler, but we know that he was a patient and industrious worker. His idleness was his way of describing his long musings, waiting the bidding of her whom God inspires—Truth, who often hides her face from the clouded eyes of man. For hours, days, weeks, he was disinclined to work. He felt no constraining impulse, his attention was relaxed or engaged upon a novel, or his seeds, or the plan of a new house, which always excited his interest. Then, apparently suddenly, whilst in one of his day-dreams, or in a fever (as at Ternate, to recall the historical episode when the theory of Natural Selection struck him), an explanation, a theory, a discovery,[69] the plan of a new book, came to him like a flash of light, and with the plan the material, the arguments, the illustrations; the words came tumbling one over the other in his brain, and as suddenly his idleness vanished, and work, eager, prolonged, unwearying, filled his days and months and years until the message was written down and the task fully accomplished. Whilst writing he referred to few books, but wrote straight on, adding paragraph to paragraph, chapter to chapter, without recasting or revision.[70] And the result was fresh, striking, original. It was a creation. The work being done, he relapsed into his busy idleness. The truth, as he saw it, seemed to come to him. Some people called him a prophet, but he was not conscious of that high calling. I do not remember him saying that he was only a messenger. Perhaps later, when he was reviewing his life, he connected his sudden inspirations with a higher source, but for their realisation he relied upon a foundation of veritable facts, facts patiently accumulated, a foundation laid broad and deep. He had the vision of the prophet allied with the wisdom of the philosopher and the calm mental detachment of the man of science. Perhaps another explanation of his genius may be found in his open-mindedness. Truth found ready access to his conscience, and always a warm welcome, and he saw with open eyes where others were stone-blind.

He belonged to our common humanity. No caste or acquired pride or unapproachable intellectualism cut him off from the people. His simple humanness made him one with us all. And his humanity was singularly comprehensive. It led him, for instance, to investigate the subject of suffering in animals. He noticed that all good men and women rightly shrank from giving pain to them, and he set himself to prove that the capacity for pain decreased as we descended the scale of life, and that poets and others were mistaken when they imputed acute suffering to the lower creation, because of the very restricted response of their nervous system. Even in the case of the human infant, he concluded that only very slight sensations are at first required, and that such only are therefore developed. The sensation of pain does not, probably, reach its maximum till the whole organism is fully developed in the adult individual. "This," he added, with that characteristic touch which made him kin to all oppressed people, "is rather comforting in view of the sufferings of so many infants needlessly sacrificed through the terrible defects of our vicious social system."

To Wallace pain was the birth-cry of a soul's advance—the stamp of rank in nature is capacity for pain. Pain, he held, was always strictly subordinated to the law of utility, and was never developed beyond what was actually needed for the protection and advance of life. This brings the sensitive soul immense relief. Our susceptibility to the higher agonies is a condition of our advance in life's pageant.

Take another instance. Amongst his numerous correspondents there were not a few who decided not to take life, for food, or science, or in war. One young man who went out with the assistance of Wallace to Trinidad and Brazil to become a naturalist, and to whom he wrote many letters[71] of direction and encouragement, gave up the work of collecting—to Wallace's sincere disappointment—and came home because he felt that it was wrong to take the lives of such wondrous and beautiful birds and insects. Another correspondent, who had joined the Navy, wrote a number of long letters to Wallace setting forth his conscientious objections to killing, arrived at after reading Wallace's books; and although Wallace endeavoured from prudential considerations to restrain him from giving up his position, he nevertheless wholly sympathised with him and in the end warmly defended him when it was necessary to do so. The sacrifice, too, of human life in dangerous employments for the purpose of financial gain, no less than the frightful slaughter of the battlefield, was abhorrent to Wallace and aroused his intensest indignation. Life to him was sacred. It had its origin in the spiritual kingdom. "We are lovers of nature, from 'bugs' up to 'humans,'" he wrote to Mr. Fred Birch.

By every means he laboured earnestly to secure an equal opportunity of leading a useful and happy life for all men and women. He championed the cause of women—of their freer life and their more active and public part in national service. He found the selective agency, which was to work for the amelioration he desired, in a higher form of sexual selection, which will be the prerogative of women; and therefore woman's position in the not distant future "will be far higher and more important than any which has been claimed for or by her in the past." When political and social rights are conceded to her on equality with men, her free choice in marriage, no longer influenced by economic and social considerations, will guide the future moral progress of the race, restore the lost equality of opportunity to every child born in our country, and secure the balance between the sexes. "It will be their (women's) special duty so to mould public opinion, through home training and social influence, as to render the women of the future the regenerators of the entire human race."

He was acutely anxious that his ideals should be realised on earth by the masses of the people. He had a large and noble vision of their future. And he had his plan for their immediate redemption—national ownership of the soil, better housing, higher wages, certainty of employment, abolition of preventable diseases, more leisure and wider education, not merely for the practical work of obtaining a livelihood but to enable them to enjoy art and literature and song. His opposition to Eugenics (to adopt the word introduced by Galton, which Wallace called jargon) sprang from his idealism and his love of the people, as well as from his scientific knowledge. On the social side he thought that Eugenics offered less chance of a much-needed improvement of environment than the social reforms which he advocated, whilst on the scientific side he believed that the attempt, with our extremely limited knowledge, to breed men and women by artificial selection was worse than folly. He feared that, as he understood it, Eugenics would perpetuate class distinctions, and postpone social reform, and afford quasi-scientific excuses for keeping people "in the positions Nature intended them to occupy," a scientific reading of the more offensive saying of those who, having plenty themselves, believe that it is for the good of the lower classes to be dependent upon others. "Clear up," he said to the present writer one day, when we drifted into a warm discussion of the teachings of Eugenists; "change the environment so that all may have an adequate opportunity of living a useful and happy life, and give woman a free choke in marriage; and when that has been going on for some generations you may be in a better position to apply whatever has been discovered about heredity and human breeding, and you may then know which are the better stocks."

"Segregation of the unfit," he remarked to an interviewer after the Eugenic Conference, at which much was unhappily said that wholly justified his caustic denunciation, "is a mere excuse for establishing a medical tyranny. And we have enough of this kind of tyranny already ... the world does not want the eugenist to set it straight.... Eugenics is simply the meddlesome interference of an arrogant scientific priestcraft."

Thus his radicalism and his so-called fads were born of his high aspirations. He was not the recluse calmly spinning theories from a bewildering chaos of observations, and building up isolated facts into the unity of a great and illuminating conception in the silence and solitude of his library, unmindful of the great world of sin and sorrow without. He could say with Darwin, "I was born a naturalist"; but we can add that his heart was on fire with love for the toiling masses. He had felt the intense joy of discovering a vast and splendid generalisation, which not only worked a complete revolution in biological science, but has also illuminated the whole field of human knowledge. Yet his greatest ambition was to improve the cruel conditions under which thousands of his fellow-creatures suffered and died, and to make their lives sweeter and happier. His mind was great enough and his heart large enough to encompass all that lies between the visible horizons of human thought and activity, and even in his old age he lived upon the topmost peaks, eagerly looking for the horizon beyond. In the words of the late Mr. Gladstone, he "was inspired with the belief that life was a great and noble calling; not a mean and grovelling thing that we are to shuffle through as we can, but an elevated and lofty destiny."

But we must not be tempted into further disquisition. As he grew older the public Press as well as his friends celebrated his birthdays. Congratulations by telegram and letter poured in upon him and gave him great pleasure. Minor poets sang special solos, or joined in the chorus. One example may be quoted:

## ALFRED RUSSEL WALLACE
### 8TH JANUARY, 1911

A little cot back'd by a wood-fring'd height,

Where sylvan Usk runs swiftly babbling by:

Here thy young eyes first look'd on earth and sky,

And all the wonders of the day and night;

O born interpreter of Nature's might,

Lord of the quiet heart and seeing eye,

Vast is our debt to thee we'll ne'er deny,

Though some may own it in their own despite.

Now after fourscore teeming years and seven,

Our hearts are jocund that we have thee still

A refuge in this world of good and ill,

When evil triumphs and our souls are riv'n;

A friend to all the friendless under heav'n;

A foe to fraud and all the lusts that kill.


O champion of the Truth, whate'er it be!

World-wand'rer over this terrestrial frame;

Twin-named with Darwin on the roll of fame;

This day we render homage unto thee;

For in thy steps o'er alien land and sea,

Where life burns fast and tropic splendours flame.

Oft have we follow'd with sincere acclaim

To mark thee unfold Nature's mystery.

For this we thank thee, yet one thing remains

Shall shrine thee deeper in the heart of man,

In ages yet to be when we are dust;

Thou hast put forth thy hand to rend our chains,

Our birthright to restore from feudal ban;

O righteous soul, magnanimous and just!

W. BRAUNSTON JONES.

Sir William Barrett, one of Wallace's oldest friends, visited him during the last year of his life, and thus describes the visit:

> In the early summer of 1913, some six months before his death, I had the pleasure of paying another visit and spending a delightful afternoon with my old friend. His health was failing, and he sat wrapped up before a fire in his study, though it was a warm day. He could not walk round his garden with me as before, but pointed to the little plot of ground in front of the French windows of his study—where he had moved some of his rarer primulas and other plants he was engaged in hybridising—and which he could just manage to visit. His eyesight and hearing seemed as good as ever, and his intellectual power was undimmed....
>
> Dr. Wallace then, pointing to the beautiful expanse of garden, woodland and sea which was visible from the large study windows, burst forth with vigorous gesticulation and flashing eyes: "Just think! All this wonderful beauty and diversity of nature results from the operation of a few simple laws. In my early unregenerate days I used to think that only material forces and natural laws were operative throughout the world. But these I now see are hopelessly inadequate to explain this mystery and wonder and variety of life. I am, as you know, absolutely convinced that behind and beyond all elementary processes there is a guiding

and directive force; a Divine power or hierarchy of powers, ever controlling these processes so that they are tending to more abundant and to higher types of life."

This led Dr. Wallace to refer to my published lecture on "Creative Thought" and express his hearty concurrence with the line of argument therein; in fact he had already sent me his views, which, with his consent, I published as a postscript to that lecture.

Then our conversation turned upon recent political events, and it was remarkable how closely he had followed, and how heartily he approved, the legislation of the Liberal Government of the day. His admiration for Mr. Lloyd George was unfeigned. "To think that I should have lived to see so earnest and democratic a Chancellor of the Exchequer!" he exclaimed, and he confidently awaited still larger measures which would raise the condition of the workers to a higher level; and nothing was more striking than his intense sympathy with every movement for the relief of poverty and the betterment of the wage-earning classes. The land question, we agreed, lay at the root of the matter, and land nationalisation the true solution. In fact, ever since I read the proof-sheets of his book on this subject, which he corrected when staying at my house in Kingstown, I have been a member of the Land Nationalisation Society, of which he was President.

Needless to say, Dr. Wallace was an ardent Home Ruler and Free Trader,[72] but on the latter question he said there should be an export duty on coal, especially the South Wales steam coal, as our supply was limited and it was essential for the prosperity of the country— and "the purchaser pays the duty," he remarked. I heartily agreed with him, and said that a small export duty *had* been placed on coal by the Conservative Government, but subsequently was removed. This he had forgotten, and when later on I sent him particulars of the duty and its yield, he replied saying that at that time he was so busy with the preparation of a book that he had overlooked the fact. He wrote most energetically on the importance of the Government

> being wise in time, and urged at least a 2s. export duty on coal.
>
> We talked about the question of a portrait of Dr. Wallace being painted and presented to the Royal Society, which had been suggested by the Rev. James Marchant, to whom Dr. Wallace referred, when talking to me, in grateful and glowing terms.—W.F.B.

Perhaps it should be added to Sir William Barrett's reminiscences that the movement which was set on foot to carry out this project was stayed by Wallace's death.

During the last years of his life his pen was seldom dry. His interest in science and in politics was fresh and keen to the closing week. He wrote "Social Environment and Moral Progress" in 1912, at the age of 90. The book had a remarkable reception. Leading articles and illustrated reviews appeared in most of the daily newspapers. The book, into which he had put his deepest thoughts and feelings upon the condition of society, was hailed as a virile and notable production from a truly great man. After this was issued, he saw another, "The Revolt of Democracy," through the press. But this did not exhaust his activities. He entered almost immediately into a contract to write a big volume upon the social order, and as a side issue to help, as is mentioned in the Introduction, in the production of an even larger book upon the writings and position of Darwin and Wallace and the theory of Natural Selection as an adequate explanation of organic evolution. Age did not seem to weaken his amazing fertility of creative thought, nor to render him less susceptible to the claims of humanity, which he faced with a noble courage. In nobility of character and in magnitude, variety and richness of mind he was amongst the foremost scientific men of the Victorian Age, and with his death that great period, which was marked by wide and illuminating generalisations and the grand style in science, came to an end.

Apart altogether, however, from his scientific position and attainments, which set him on high, he was a noble example of brave, resolute, and hopeful endeavour, maintained without faltering to the end of a long life. And this is not the least valuable part of his legacy to the race.

When Henslow died, Huxley wrote to Hooker: "He had intellect to comprehend his highest duty distinctly, and force of character to do it; which of us dare ask for a higher summary of his life than that? For such a man there can be no fear in facing the great unknown; his life has been one long experience of the substantial justice of the laws by which this world is governed, and he will calmly trust to them still as he lays his head down for his long sleep." Let that also stand as the estimate of Wallace by his contemporaries, an estimate which we believe posterity will confirm. And to it we may add that death, which came to him in his sleep as a gentle deliverer, opened the door into the larger and fuller life into which he tried to penetrate and in which he firmly believed. If that faith be founded in truth, Darwin and Wallace, yonder as here, are united evermore.

I am writing these concluding words on the second anniversary of his death. Before me there lies the telegram which brought me the sad news that he had "passed away very peacefully at 9.25 a.m., without regaining consciousness." He was in his ninety-first year. It was suggested that he should be buried in Westminster Abbey, beside Charles Darwin, but Mrs. Wallace and the family, expressing his own wishes as well as theirs, did not desire it. On Monday, November 10th, he was laid to rest with touching simplicity in the little cemetery of Broadstone, on a pine-clad hill swept by ocean breezes. He was followed on his last earthly journey by his son and daughter, by Miss Mitten, his sister-in-law, and by the present writer. Mrs. Wallace, being an invalid, was unable to attend. The funeral service was conducted by the Bishop of Salisbury (Dr. Ridgeway), and among the official representatives were Prof. Raphael Meldola and Prof. E.B. Poulton representing the Royal Society; the latter and Dr. Scott representing the Linnean Society, and Mr. Joseph Hyder the Land Nationalisation Society. A singularly appropriate monument, consisting of a fossil tree-trunk from the Portland beds, has been erected over his grave upon a base of Purbeck stone, which bears the following inscription:

ALFRED RUSSEL WALLACE, O.M.
Born Jan. 8th, 1823, Died Nov. 7th, 1913

A year later, on the 10th of December, 1914, his widow died after a long illness, and was buried in the same grave. She was the eldest daughter of Mr. William Mitten, of Hurstpierpoint, an enthusiastic botanist, and in no mean degree she inherited her father's love of wild flowers and of the beautiful in nature. It was this similarity of tastes which led to her close intimacy and subsequent marriage, in 1866, with Wallace. Their married life was an exceedingly happy one. She was able to help him in his scientific labours, and she provided that atmosphere in the home life which enabled him to devote himself to his many-sided enterprises. And nothing would give him more joy than to know that this book is dedicated to her memory.

## THE GRAVE OF ALFRED RUSSEL AND ANNIE WALLACE

Soon after Wallace's death a Committee was formed (with Prof. Poulton as Chairman and Prof. Meldola as Treasurer) to erect a memorial, and the following petition was sent to the Dean and Chapter of Westminster Abbey:

> We, the undersigned, earnestly desiring a suitable national memorial to the late Alfred Russel Wallace, and believing that no position would be so appropriate as Westminster Abbey, the burial-place of his illustrious fellow-worker Charles Darwin, petition the Right Reverend the Dean and Chapter for permission to place a medallion in Westminster Abbey. We further guarantee, if the medallion be accepted, to pay the Abbey fees of £200.
>
> ARCH. GEIKIE
>
> WILLIAM CROOKES
>
> A.B. KEMPE
>
> E. RAY LANKESTER
>
> D.H. SCOTT

D. PRAIN

A.E. SHIPLEY

RAPHAEL MELDOLA

P.A. MACMAHON

JOHN W. JUDD

OLIVER J. LODGE

E.B. POULTON

A. STRAHAN

H.H. TURNER

J. LARMOR

W. RAMSAY

SILVANUS P. THOMPSON

JOHN PERRY

JAMES MARCHANT (Hon. Sec.)

To which the Dean replied:

*The Deanery, Westminster, S.W. December 2, 1913.*

Dear Mr. Marchant,—I have pleasure in informing you that I presented your petition at our Chapter meeting this morning, and a glad and unanimous assent was accorded to it.

I should be glad later on to be informed as to the artist you are employing; and probably it would be as well for him and you and some members of the Royal Society to meet me and the Chapter and confer together upon the most suitable and artistic arrangement or rearrangement of the medallions of the great men of science of the nineteenth century.

Nothing could have been more satisfactory or impressive than the document with which you furnished me this morning. I hope to get it specially framed.—Yours sincerely,

HERBERT E. RYLE.

Mr. Bruce-Joy, who had made an excellent medallion of Dr. Wallace during his lifetime, accepted the commission to fashion the medallion for Westminster Abbey, and it was unveiled, by a happy but undesigned coincidence, on All Souls' Day, November 1 1915, together with medallions to the memory of Sir Joseph Hooker and Lord Lister. In the course of his sermon, the Dean said—and with these words we may well conclude this book:

"To-day there are uncovered to the public view, in the North Aisle of the Choir, three memorials to men who, I believe, will always be ranked among the most eminent scientists of the last century. They passed away, one in 1911, one in 1912, and one in 1913. They were all men of singularly modest character. As is so often observable in true greatness, there was in them an entire absence of that vanity and self-advertisement which are not infrequent with smaller minds. It is the little men who push themselves into prominence through dread of being overlooked. It is the great men who work for the work's sake without regard to recognition, and who, as we might say, achieve greatness in spite of themselves.

THE WALLACE AND DARWIN MEDALLIONS IN THE NORTH AISLE OF THE CHOIR OF WESTMINSTER ABBEY

"Alfred Russel Wallace was a most famous naturalist and zoologist. He arrived by a flash of genius at the same conclusions which Darwin had reached after sixteen years of most minute toil and careful observation.... It was a unique example of the almost exact concurrence of two great minds working upon the same subject, though in different parts of the world, without collusion and without rivalry.... Between Darwin and Wallace goodwill and friendship were never interrupted. Wallace's life was spent in the pursuit of various objects of intellectual and philosophical interest, over which I need not here linger. All will agree that it is fitting his medallion should be placed next to that of Darwin, with whose great name his own will ever be linked in the worlds of thought and science.

"All will acknowledge the propriety of these three great names being honoured in this Abbey Church, even though it be, to use Wordsworth's phrase, already

'Filled with mementoes, satiate with its part

Of grateful England's overflowing dead.'

"These are three men whose lifework it was to utilise and promote scientific discovery for the preservation and betterment of the human race."

# APPENDIX

## LISTS OF WALLACE'S WRITINGS

### I.—BOOKS

| Date | Title |
| --- | --- |
| 1853 | "Palm Trees on the Amazon" |
| 1853 | "A Narrative of Travels on the Amazon and Rio Negro." New Edition in "The Minerva Library," 1889 |
| 1866 | "The Scientific Aspect of the Supernatural" |
| 1869 | "The Malay Archipelago," 2 vols. Tenth Edition, 1 vol., 1890 |
| 1870 | "Contributions to the Theory of Natural Selection." Republished, with "Tropical Nature," 1891 |
| 1874 | "Miracles and Modern Spiritualism." Revised Edition, 1896 |
| 1876 | "The Geographical Distribution of Animals," 2 vols. |
| 1878 | "Tropical Nature and other Essays." Printed in 1 vol. with "Natural Selection," 1891 |
| 1879 | "Australasia." "Stanford's Compendium of Geography and Travel." (New issue, 1893) |
| 1880 | "Island Life." Revised Edition, 1895 |
| 1882 | "Land Nationalisation" |
| 1885 | "Bad Times" |
| 1889 | "Darwinism." 3rd Edition, 1901 |
| 1898 | "The Wonderful Century." New Edition, 1903 |
| 1900 | "Studies, Scientific and Social" |
| 1901 | "The Wonderful Century Reader" |
| 1901 | "Vaccination a Delusion" |
| 1903 | "Man's Place in the Universe." New Edition, 1904. Cheap 1s. Edition, 1912 |

| | |
|---|---|
| 1905 | "My Life," 2 vols. New Edition, 1 vol., 1908 |
| 1907 | "Is Mars Habitable?" |
| 1908 | "Notes of a Botanist on the Amazon and Andes," by Richard Spruce. Edited by A.R. Wallace |
| 1910 | "The World of Life" |
| 1913 | "Social Environment and Moral Progress" |
| 1913 | "The Revolt of Democracy" |

## II.—ARTICLES, PAPERS, REVIEWS, ETC.

*The articles marked with an asterisk were republished in Wallace's "Studies, Scientific and Social."*

| | DATE | PERIODICAL OR SOCIETY | SUBJECT |
|---|---|---|---|
| | 1850 | Proc. Zool. Soc., Lond. | On the Umbrella Bird |
| | 1852 | " " | Monkeys of the Amazon |
| | 1852-3 | Trans. Entomol. Soc. | On the Habits of the Butterflies of the Amazon Valley |
| | 1853 | Zoologist | On the Habits of the Hesperidæ |
| | 1853 | Proc. Zool. Soc., Lond. | On some Fishes allied to Gymnotus |
| June 6 | 1853 | Entomolog. Soc. | On the Insects used for Food by the Indians of the Amazon |

| June 13 | 1853 | Royal Geograph. Soc. | The Rio Negro |
|---|---|---|---|
| | 1854-5 | Zoologist | Letters from Singapore and Borneo |
| | 1854-6 | Trans. Entomol. | Description of a New Species of |
| | | Soc. | Ornithoptera |
| | 1855 | Annals and Mag. | On the Ornithology of Malacca |
| | | of Nat. Hist. | |
| | 1855 | Journ. Bot. | Botany of Malacca |
| | 1855 | Zoologist | The Entomology of Malacca |
| Sept. | 1855 | Annals and Mag. | On the Law which has regulated |
| | | of Nat. Hist. | the Introduction of New Species |
| | 1856 | " " | Some Account of an Infant |
| | | | Orang-Outang |
| | 1856 | " " | On the Orang-Outang or Mias of |
| | | | Borneo |
| Dec. | 1856 | " " | On the Habits of the Orang-Outang |
| | | | of Borneo |
| | 1856 | " " | Attempts at a Natural Arrangement |
| | | | of Birds |
| Nov. 22 | 1856 | Chambers's Journ. | A New Kind of Baby |
| | 1856 | Journ. Bot. | On the Bamboo and Durian of Borneo |
| | 1856 | Zoologist | Observations on the Zoology of |

|  |  |  |  | Borneo |
|---|---|---|---|---|
|  | 1856-8 | Trans. Entomol. |  | On the Habits, etc., of a Species |
|  |  | Soc. |  | of Ornithoptera inhabiting the |
|  |  |  |  | Aru Islands |
|  | 1856-9 | " " |  | Letters from Aru Islands and from |
|  |  |  |  | Batchian |
| Dec. | 1857 | Annals and Mag. |  | Natural History of the Aru Islands |
|  |  | of Nat. Hist. |  |  |
|  | 1857 | " " |  | On the Great Bird of Paradise |
|  | 1857 | Proc. Geograph. |  | Notes of a Journey up the Sadong |
|  |  | Soc. |  | River |
|  | 1858 | " " |  | On the Aru Islands |
|  | 1858 | Zoologist |  | Note on the Theory of Permanent |
|  |  | " " |  | and Geographical Varieties |
|  | 1858 | " " |  | On the Entomology of the Aru |
|  |  |  |  | Islands |
|  | 1858-61 | Trans. Entomol. |  | Note on the Sexual Differences in |
|  |  | Soc. |  | the Genus Lomaptera |
|  | 1859 | Annals and Mag. |  | Correction of an Important Error |
|  |  | of Nat. Hist. |  | affecting the Classification of |
|  |  |  |  | the _Psittacidæ_ |

|      | 1859 | Proc, Linn. Soc. | On the Tendency of Varieties to |
|------|------|------------------|------|
|      |      | (iii. 45)        | Depart Indefinitely from the |
|      |      |                  | Original Type[73] |
| Oct. | 1859 | Ibis             | Geographical Distribution of Birds |
| Dec. | 1859 | Entomolog. Soc.  | Note on the Habits of Scolytidæ and |
|      |      |                  | Bostrichidæ |
|      | 1860 | Journ. Geograph. | Notes of a Voyage to New Guinea |
|      |      | Soc.             |      |
|      | 1860 | Ibis             | The Ornithology of North Celebes |
|      | 1860 | Proc. Zool, Soc., | Notes on Semioptera wallacii |
|      |      | Lond.            |      |
|      | 1860 | Proc. Linn. Soc. | Zoological Geography of Malay |
|      |      | (iv. 172)        | Archipelago |
|      | 1861 | Ibis             | On the Ornithology of Ceram and |
|      |      |                  | Waigiou |
|      | 1861 | "                | Notes on the Ornithology of Timor |
|      | 1862 | Proc. and Journ. | On the Trade between the Eastern |
|      |      | Geogr. Soc.      | Archipelago and New Guinea |
|      |      |                  | and its Islands |
|      | 1862 | Proc. Zool. Soc., | List of Birds from the Sula Islands |

| | | | |
|---|---|---|---|
| | | Lond. | |
| | 1862 | Ibis | On some New Birds from the Northern |
| | | | Moluccas |
| | 1862 | Proc. Zool. Soc., | Narrative of Search after Birds of |
| | | Lond. | Paradise |
| | 1862 | " | On some New and Rare Birds from New |
| | | | Guinea |
| | 1862 | " | Description of Three New Species |
| | | | of _Pitta_ from the Moluccas |
| | 1863 | Annals and Mag. | On the Proposed Change in Name of |
| | | of Nat. Hist. | _Gracula pectoralis_ |
| | 1863 | Entomol. Journ. | Notes on the Genus _Iphias_ |
| | 1863 | Ibis | Note on _Corvus senex_ and _Corvus |
| | | | fuscicapillus_ |
| | 1863 | " | Notes on the Fruit-Pigeons of Genus |
| | | | _Treron_ |
| | 1863 | Intellectual | The Bucerotidæ, or Hornbills |
| | | Observer | |
| | 1863 | Proc. Zool, Soc. | List of Birds collected on Island |
| | | Lond. | of Bouru |

| April | 1863 | Zoologist | Who are the Humming-Bird's |
|---|---|---|---|
| | | | Relations? |
| June | 1863 | Royal Geograph. | Physical Geography of the Malay |
| | | Soc. | Archipelago |
| | 1863 | Proc, Zool. Soc., | On the Identification of _Hirundo |
| | | Lond. | esculenta_, Linn. |
| | 1863 | " | List of Birds inhabiting the Islands |
| | | | of Timor, Flores and Lombok |
| | 1863 | Annals and Mag. | On the Rev. S. Haughton's Paper on |
| | | of Nat. Hist. | the Bee's Cell and the Origin of |
| | | | Species |
| Jan. 1 | | Nat. Hist. Rev. | Some Anomalies in Zoological and |
| | | | Botanical Geography |
| Jan. 7 | 1864 | Edinburgh New | Ditto |
| | | Journ. (Philos.) | |
| | 1864 | Proc. Zool. Soc., | Parrots of the Malayan Region |
| | | Lond. | |
| | 1864 | Anthropol. Soc. | The Origin of Human Races and the |
| | | Journ. | Antiquity of Man deduced from |
| | | | Natural Selection |

- 196 -

|  | 1864 | Proc. Entom. Soc. | Effect of Locality in producing |
|  |  | and Zoologist | Change of Form in Insects |
|  | 1864 | Proc. Entom. Soc. | Views on Polymorphism |
|  | 1864 | Ibis | Remarks on the Value of |
|  |  |  | Osteological Characters in the |
|  |  |  | Classification of Birds |
|  | 1864 | " | Remarks on the Habits, |
|  |  |  | Distribution, etc., of the Genus |
|  |  |  | _Pitta_ |
|  | 1864 | " | Note on _Astur griseiceps_ |
|  | 1864 | Nat. Hist. Rev. | Bone Caves in Borneo |
|  | 1865 | Proc. Zool. Soc., | List of the Land Shells collected |
|  |  | Lond. | by Mr. Wallace in the Malay |
|  |  |  | Archipelago |
| Jan. | 1865 | Trans. Ethnolog. | On the Progress of Civilisation in |
|  |  | Soc. | North Celebes |
| Jan. | 1865 | " | On the Varieties of Man in the |
|  |  |  | Malay Archipelago |
|  | 1865 | Proc. Zool. Soc., | Descriptions of New Birds from the |
|  |  | Lond. | Malay Archipelago |
| June 17 | 1865 | Reader | How to Civilise Savages* |
| Oct. | 1865 | Ibis | Pigeons of the Malay Archipelago |

|  | 1866 | Trans. Linn. Soc. | On the Phenomena of Variation and |
|  |  | (xxv.) (Abstract | Geographical Distribution as |
|  |  | in Reader, April, | illustrated by Papilionidæ of |
|  |  | 1864) | the Malayan Region |
|  | 1866 | Proc. Zoo. Soc., | List of Lepidoptera collected by |
|  |  | Lond. | Swinton at Takow, Formosa |
|  | 1866 | Proc. Entomol. } | Exposition of the Theory of |
|  |  | Soc. } | Mimicry as explaining Anomalies |
|  | 1867 | Zoologist } | of Sexual Variation |
|  | 1867 | Intellectual | The Philosophy of Birds' Nests |
|  |  | Observer |  |
| Jan. | 1867 | Quarterly Journ. | Ice-Marks in North Wales |
|  |  | of Sci. |  |
| April | 1867 | " | The Polynesians and their |
|  |  |  | Migrations* |
| July | 1867 | Westminster Rev. | Mimicry and other Protective |
|  |  |  | Resemblances among Animals |
| Sept. | 1867 | Science Gossip | Disguises of Insects |
| Oct. | 1867 | Quarterly Journ. | Creation by Law |
|  |  | of Sci. |  |
|  | 1867 | Proc. Entomol. } |  |
|  |  | Soc. } | A Catalogue of the Cetoniidæ of |

|  | 1868 | Trans. Entomol. } | the Malayan Archipelago, etc. |
|---|---|---|---|
|  |  | Soc. } |  |
| Jan. 7 | 1868 | Ibis | Raptorial Birds of the Malay Archipelago |
|  | 1868 | Trans. Entomol. Soc. | On the Pieridæ of the Indian and Australian Regions |
|  | 1868 | — | The Limits of Natural Selection applied to Man* |
|  | 1869 | Trans. Entomol. Soc. | Note on the Localities given in the "Longicornia Malayana" |
|  | 1869 | Journ. of Travel and Nat. Hist. | A Theory of Birds' Nests |
| April | 1869 | Quarterly Rev. | Reviews of Lyell's "Principles of Geology" (entitled "Geological Climates and Origin of Species") |
|  | 1869 | Macmillan's Mag. | Museums for the People* |
|  | 1869 | Trans. Entomol. Soc. | Notes on Eastern Butterflies (3 Parts) |
|  | 1870 | Brit. Association Report | On a Diagram of the Earth's Eccentricity, etc. |
| March | 1871 | Academy | Review of Darwin's "Descent of Man" |

| | | | |
|---|---|---|---|
| May 23 | 1871 | Entomolog. Soc. | Address on Insular Faunas, etc. |
| | 1871 | " | The Beetles of Madeira and their Teachings* |
| Nov. | 1871 | —— | Reply to Mr. Hampden's Charges |
| | 1873 | Journ. Linnean Soc. | Introduction to F. Smith's Catalogue of Aculeate Hymenoptera, etc. |
| Jan. 4 | 1873 | Times | Spiritualism and Science |
| April | 1873 | Macmillan's Mag. | Disestablishment and Disendowment, with a Proposal for a really National Church of England* |
| Sept. 16 | 1873 | Daily News | Coal a National Trust* |
| Dec. | 1873 | Contemp. Rev. | Limitation of State Functions in the Administration of Justice* |
| Jan. 17 | 1874 | Academy | Reviews of Mivart's "Man and Apes" and A.J. Mott's "Origin of Savage Life" |
| April | 1874 | —— | Review of W. Marshall's "Phrenologist amongst the Todas" |
| April | 1874 | —— | Review of G. St. Clair's "Darwinism and Design" |

|  | 1874 | Ibis | On the Arrangement of the Families constituting the Order Passeres |
|---|---|---|---|
| May | 1876 | Academy | Review of Mivart's "Lessons from Nature" |
|  | 1877 | Proc. Geograph. Soc. | The Comparative Antiquity of Continents |
| July | 1877 | Quarterly Journ. of Sci. | Review of Carpenter's "Mesmerism and Spiritualism," etc. |
| Sept. and Oct. | 1877 | Macmillan's Mag. | The Colours of Animals and Plants |
| Nov. | 1877 | Fraser's Mag. | The Curiosities of Credulity |
| Dec. | 1877 | Fortnightly Rev. | Humming-Birds |
| Dec. | 1877} | Athenæum | {Correspondence with W.B. |
| Jan. | 1878} | " | { Carpenter on Spiritualism |
| Nov. | 1878 | Fortnightly Rev. | Epping Forest, and How to Deal with it |
| Feb. | 1879 | Contemp. Rev. | New Guinea and its Inhabitants |
| April | 1879 | Academy | Review of Haeckel's "Evolution of Man" |
| July | 1879 | Nineteenth Cent. | Reciprocity: A Few Words in |

| | | | | |
|---|---|---|---|---|
| | | | | Reply to Mr. Lowe* |
| July | 1879 | Quarterly Rev. | | Glacial Epochs and Warm Polar |
| | | | | Climates |
| Jan. | 1880 | Nineteenth Cent. | | The Origin of Species and |
| | | | | Genera* |
| Oct. | 1880 | Academy | | Review of A.H. Swinton's |
| | | | | "Insect Variety" |
| Nov. | 1880 | Contemp. Rev. | | How to Nationalise the Land* |
| Dec. 4 | 1880 | Academy | | Review of Seebohm's "Siberia In |
| | | | | Europe" |
| | 1881 | Rugby Nat. Hist. | | Abstract of Four Lectures on |
| | | Soc. Rept. | | the Natural History of |
| | | | | Islands |
| Dec. | 1881 | Contemp. Rev. | | Monkeys: Their Affinities and |
| | | | | Distribution* |
| Aug. and | 1883 | Macmillan's Mag. | | The Why and How of Land |
| Sept. | | | | Nationalisation* |
| March | 1884 | Christn. Socialist | | The Morality of Interest—The |
| | | | | Tyranny of Capital |
| | 1886 | Claims of Labour | | The Depression of Trade* |
| | | Lectures | | |
| Mar. 5 | 1887 | Banner of Light | | Letter "_In re_ Mrs. Ross |
| | | | | (Washington, D.C.)" |

| Mar. 17 | 1887 | Independ. Rev. | Review of E.D. Cope's "Origin |
|---|---|---|---|
| | | | of the Fittest" |
| | 1887 | Nation | " |
| Oct. | 1887 | Fortnightly Rev. | American Museums* |
| | 1888 | —— | The Action of Natural Selection |
| | | | in producing Old Age, Decay |
| | | | and Death |
| June | 1889 | Land Nationalisation Soc. | Address |
| Sept. | 1890 | Fortnightly Rev. | Progress without Poverty (Human |
| | | | Selection)* |
| Oct. | 1891 | " | English and American Flowers* |
| Dec. | 1891 | " | Flowers and Forests of the Far |
| | | | West* |
| Jan. | 1892 | Arena | Human Progress, Past and |
| | | | Future* |
| | 1892 | Address to L.N.S. | Herbert Spencer on the Land |
| | | | Question* |
| Aug. | 1892 | Nineteenth Cent. | Why I Voted for Mr. Gladstone |
| Aug. and Dec. | 1892 | Natural Sci. | The Permanence of Great Ocean Basins* |
| Nov. | 1892 | Fortnightly Rev. | Our Molten Globe* |

| Dec. | 1892 | Natural Sci. | Note on Sexual Selection |
|---|---|---|---|
| Feb. | 1893 | Nineteenth Cent. | Inaccessible Valleys* |
| Mar. and Apr. | 1893 | Arena | The Social Quagmire and the Way Out of it* |
| Apr. and May | 1893 | Fortnightly Rev. | Are Individually Acquired Characters Inherited?* |
| Nov. | 1893 | " | The Ice Age and its Work* |
| Dec. | 1893 | " | Erratic Blocks, etc. Lake Basins* |
| | 1893 | Arena | The Bacon-Shakespeare Case |
| April 9 | 1894 | Land Nationalisation Soc. | Address on Parish Councils |
| June | 1894 | Natural Sci. | The Palearctic and Nearctic Regions compared as regards Families and Genera of Mammalia and Birds |
| June | 1894 | Contemp. Rev. | How to Preserve the House of Lords* |
| July | 1894 | Land and Labour | Review of F.W. Hayes' "Great Revolution of 1905" |
| Sept. | 1894 | Natural Sci. | The Rev. G. Henslow on Natural Selection* |
| | 1894 | Smithsonian Rep. | Method of Organic Evolution |

- 204 -

| | | | |
|---|---|---|---|
| Oct. | 1894 | Nineteenth Cent. | A Counsel of Perfection for Sabbatarians* |
| | 1894 | Vox Clamantium | Economic and Social Justice* |
| Feb. and March | 1895 | Fortnightly Rev. | Method of Organic Evolution* |
| Oct. | 1895 | " | Expressiveness of Speech or Mouth-Gesture as a Factor in the Origin of Language* |
| | 1895 | Agnostic Annual | Why Live a Moral Life?* |
| May | 1896 | Contemp. Rev. | How Best to Model the Earth* |
| July 25 | 1896 | Labour Leader | Letter on International Labour Congress |
| Aug. | 1896 | Fortnightly Rev. | The Gorge of the Aar and its Teaching* |
| Dec. | 1896 | Journ. Linn. Soc. (v. 25) | The Problem of Utility: Are Specific Characters always or generally Useful? |
| March | 1897 | Natural Sci. | Problem of Instinct* |
| | 1897 | "Forecasts of Coming Century" | Re-occupation of Land, Solution of the Unemployed Problem* |
| March 20 | 1898 | Lancet | Letter on Vaccination |

| | | | |
|---|---|---|---|
| May 9 | 1898 | Shrewsbury Chron. | Letter to Dr. Bond and A.K.W. |
| | | | on Vaccination |
| June 16, 21, 25, Aug. 15 | 1898 | Echo | " |
| Sept. 1 | 1898 | The Eagle and the Serpent | Darwinism and Nietzscheism in Sociology |
| | 1898 | Printed for private circulation | Justice not Charity (Address to International Congress of Spiritualists, London, June, 1898)* |
| Dec. 31 | 1898 | Academy | Paper Money as a Standard of Value* |
| Feb., March, April | 1899 | Journ. Soc. Psychical Res. | Letters on Mr. Podmore _re_ Clairvoyance, etc. |
| May | 1899 | L' Humanité Nouvelle | The Causes of War and the Remedies* |
| Nov. 18 | 1899 | Clarion | Letter on the Transvaal War |
| | 1899 | N.Y. Independent | White Men in the Tropics* |
| | 1900 | N.Y. Sun | Evolution |
| Nov. | 1900 | N.Y. Journ. | Social Evolution in the Twentieth Century: An Anticipation |
| | 1900 | ——— | Ralahine and its Teachings* |

|  |  |  |  | True Individualism the Essential Preliminary of a Real Social Advance* |
|---|---|---|---|---|
|  | 1901 | Morning Leader | | An Appreciation of the Past Century |
| Jan. 17 | 1903 | Black and White | | Relations with Darwin |
| March | 1903 | Fortnightly Rev. | | Man's Place in the Universe |
| Sept. | 1903 | " | | Man's Place in the Universe. Reply to Critics |
| Oct. | 1903 | Academy | | The Wonderful Century. Reply to Dr. Saleeby |
| Nov. 12 | 1903 | Daily Mail | | Does Man Exist in Other Worlds? Reply to Critics |
| Jan. 1 | 1904 | Clarion | | Anticipations for the Immediate Future, Written for the _Berliner Lokalanzeiger_, and refused |
| Feb., April | 1904 | Fortnightly Rev. | | An Unpublished Poem by E.A. Poe, "Leonainie" |
| Apr., May | 1904 | Independent Rev. | | Birds of Paradise in the Arabian Nights |
|  | 1904 | Anti-Vaccination League | | Summary of the Proofs that Vaccination does not Prevent Small-pox, but really |

|  |  |  |  | Increases it |
|---|---|---|---|---|
|  | 1904 | Labour Annual |  | Inefficiency of Strikes |
|  | 1904 | Clarion |  | Letter on Opposition to |
|  |  |  |  | Military Expenditure |
|  |  | Vaccination |  | Letter on Inconsistency of the |
|  |  | Inquirer |  | Government on Vaccination |
| Oct. 27 | 1906 | Daily News |  | Why Not British Guiana? Five |
|  |  |  |  | Acres for 2s. 6d. |
| Nov. | 1906 | Independent Rev. |  | The Native Problem in South |
|  |  |  |  | Africa and Elsewhere |
| Jan. | 1907 | Fortnightly Rev. |  | Personal Suffrage, a Rational |
|  |  |  |  | System of Representation and |
|  |  |  |  | Election |
| Feb. | 1907 | " |  | A New House of Lords |
|  | 1907 | Harmsworth's "History |  | How Life became Possible on the |
|  |  | of the World" |  | Earth |
| Sept. 13 | 1907 | Public Opinion |  | Letter on Sir W. Ramsay's |
|  |  |  |  | Theory: Did Man reach his |
|  |  |  |  | Highest Development in the |
|  |  |  |  | Past? |
| Jan. 1 | 1908 | N.Y. World |  | Cable on Advance in Science in |
|  |  |  |  | 1907 |

| Jan. 18 | 1908 | Outlook | Letter on Woman |
| Jan. | 1908 | Fortnightly Rev. | Evolution and Character |
| June and July | 1908 | Socialist Rev. | The Remedy for Unemployment |
| July | 1908 | Times | Letter on the First Paper on Natural Selection |
| July | 1908 | Delineator | Are the Dead Alive? |
| Aug. 14 | 1908 | Public Opinion | Is it Peace or War? A Reply |
| Aug. | 1908 | Contemp. Rev. | Present Position of Darwinism |
| Sept. | 1908 | New Age | Letter on Nationalisation, not Purchase, of Railways |
| Dec. | 1908 | Contemp. Rev. | Darwinism _v._ Wallaceism |
| Christ-mas | 1908 | Christian Commonwealth | On the Abolition of Want |
| Jan. 22 | 1909 | Royal Institution | The World of Life, as Visualised, etc., by Darwinism |
| Feb. | 1909 | Clarion pamphlet (? Socialist Rev.) | The Remedy for Unemployment |
| Feb. 6 | 1909 | Daily News | Flying Machines in War |
| Feb. 12 | 1909 | Daily Mail | Charles Darwin (Centenary) |
| Feb. 12 | 1909 | Clarion | The Centenary of Darwin |
| March | 1909 | Fortnightly Rev. | The World of Life (revised Lecture) |
| April 8 | 1909 | Daily News | Letter on Aerial Fleets |

| | | | |
|---|---|---|---|
| April 8 | 1910 | " | Man in the Universe |
| Oct. 14 | 1910 | Public Opinion | A New Era in Public Opinion |
| Jan. 25 | 1912 | Daily Chronicle | Letter on the Insurance Act |
| Aug. 9 | 1912 | Daily News | A Policy of Defence |
| Sept. | 1912 | ——— | The Nature and Origin of Life |

## III.—LETTERS, REVIEWS, ETC., IN "NATURE"

| VOL. | PAGE | DATE | SUBJECT |
|---|---|---|---|
| I. | 105 | 1869 | Origin of Species Controversy |
| " | 132 | " | " " " |
| " | 288, 315 | 1870 | Government Aid to Science |
| " | 399, 452 | " | Measurement of Geological Time |
| " | 501 | " | Hereditary Genius |
| II. | 82 | " | Pettigrew's "Handy Book of Bees" |
| " | 234 | " | A Twelve-wired Bird of Paradise |
| " | 350 | " | Early History of Mankind |
| " | 465 | " | Speech on the Arrangement of Specimens |
| | | " | in a Natural History Museum (British |
| | | " | Association) |
| " | 510 | " | Glaciation of Brazil |
| III. | 8, 49 | " | Man and Natural Selection |
| " | 85, 107 | " | " " " |
| " | 165 | " | Mimicry versus Hybridity |

| "    | 182      | 1871 | Leroy's "Intelligence and Perfectibility of Animals" |
| "    | 309      | "    | Theory of Glacial Motion |
| "    | 329      | "    | Duncan's "Metamorphoses of Insects" |
| "    | 385      | "    | Dr. Bevan's "Honey Bee" |
| "    | 435      | "    | Anniversary Address at the Entomological |
|      |          | "    | Society |
| "    | 466      | "    | Sharpe's Monograph of the Alcedinidæ |
| IV.  | 22       | "    | Staveley's "British Insects" |
| "    | 178      | "    | Dr. Bastian's Work on the Origin of Life |
| "    | 181      | "    | H. Howorth's Views on Darwinism |
| "    | 221      | "    | " " " |
| "    | 222      | "    | Recent Neologisms |
| "    | 282      | "    | Canon Kingsley's "At Last" |
| V.   | 350      | 1872 | The Origin of Insects |
| "    | 363      | "    | Ethnology and Spiritualism |
| VI.  | 237      | "    | The Last Attack on Darwinism (Reviews) |
| "    | 284, 299 | "    | Bastian's "Beginnings of Life" |
| "    | 328      | "    | Ocean Circulation |
| "    | 407      | "    | Speech on Diversity of Evolution (British Association) |
| "    | 469      | "    | Houzeau's "Faculties of Man and Animals" |
| VII. | 68       | "    | Misleading Cyclopædias |
| "    | 277      | 1873 | Modern Applications of the Doctrine of Natural Selection (Reviews) |
| "    | 303      | "    | Inherited Feeling |

| "     | 337      | "    | J.T. Moggridge's "Harvesting Ants and Trapdoor Spiders" |
|-------|----------|------|---------------------------------------------------------|
| "     | 461      | "    | Cave Deposits of Borneo |
| VIII. | 5        | 1873 | Natural History Collections in the East India Museum |
| "     | 65, 302  | "    | Perception and Instinct In the Lower Animals |
| "     | 358      | "    | Dr. Page's Textbook on Physical Geography |
| "     | 429      | "    | Works on African Travel (Reviews) |
| "     | 462      | "    | Lyell's "Antiquity of Man" |
| IX.   | 102      | "    | Dr. Meyer's Exploration of New Guinea |
| "     | 218      | 1874 | Belt's "Naturalist in Nicaragua " |
| "     | 258      | "    | David Sharp's "Zoological Nomenclature" |
| "     | 301, 403 | "    | Animal Locomotion |
| X.    | 459      | "    | Migration of Birds |
| "     | 502      | "    | Automatism of Animals |
| XII.  | 83       | 1875 | Lawson's "New Guinea" |
| XIV.  | 403      | 1876 | Opening Address in Biology Section, British Association |
| "     | 473      | "    | Erratum in Address to Biology Section, British Association |
| "     | 24       | "    | Reply to Reviewers of "Geographical Distribution of Animals" |
| "     | 174      | "    | "Races of Men" |
| "     | 274      | 1877 | Glacial Drift in California |
| "     | 431      | "    | The "Hog-wallows" of California |

| | | | |
|---|---|---|---|
| XVI. | 548 | " | Zoological Relations of Madagascar and Africa |
| " | 44 | " | The Radiometer and its Lessons |
| XVII. | 8 | " | Mr. Wallace and Reichenbach's Odyle |
| " | 44 | " | The Radiometer and its Lessons |
| " | 45 | " | Bees Killed by Tritoma |
| " | 100 | " | The Comparative Richness of Faunas and Floras tested Numerically |
| " | 101 | " | Mr. Crookes and Eva Fay |
| " | 182 | 1878 | Northern Affinities of Chilian Insects |
| XVIII. | 193 | " | A Twenty Years' Error in the Geography of Australia |
| XIX. | 4 | " | Remarkable Local Colour-Variation in Lizards |
| " | 121, 244 | " | The Formation of Mountains |
| " | 289 | 1879 | " " " |
| " | 477 | " | Organisation and Intelligence |
| " | 501, 581 | " | Grant Allen's "Colour Sense" |
| " | 582 | " | Did Flowers Exist during the Carboniferous Epoch |
| XX. | 141 | " | Butler's "Evolution, Old and New" |
| " | 501 | " | McCook's "Agricultural Ants of Texas" |
| " | 625 | " | Reply to Reviewers of Wallace's "Australasia " |
| XXI. | 562 | 1880 | Reply to Everett on Wallace's "Australasia" |
| XXII. | 141 | " | Two Darwinian Essays |
| XXIII. | 124, 217, | " | Geological Climates |

|  | 266 |  |  |
|---|---|---|---|
| " | 152, 175 | " | New Guinea |
| " | 169 | " | Climates of Vancouver Island and Bournemouth |
| " |  | " |  |
| " | 195 | " | Correction of an Error in "Island Life" |
| XXIV. | 242 | 1881 | Tyler's "Anthropology" |
| XXIV. | 437 | 1881 | Weismann's "Studies in the Theory of Descent" |
| XXV. | 3 | " | Carl Bock's "Head-Hunters of Borneo" |
| " | 381 | 1882 | Grant Allen's "Vignettes from Nature" |
| " | 407 | " | Houseman's "Story of Our Museum " |
| XXVI. | 52 | " | Weismann's "Studies in the Theory of Descent" |
| " | 86 | " | Müller's "Difficult Cases of Mimicry" |
| XXVII. | 481 | 1883 | " " " |
| " | 482 | " | On the Value of the Neo-arctic as One of the Primary Zoological Regions |
| XXVIII. | 293 | " | W.F. White's "Ants and their Ways" |
| XXXI. | 552 | 1885 | Colours of Arctic Animals |
| XXXII. | 218 | " | H.O. Forbes's "A Naturalist's Wanderings in the Eastern Archipelago" |
| XXXIII. | 170 | 1886 | Victor Hehn's "Wanderings of Plants and Animals" |
| XXXIV. | 333 | " | H.S. Gorham's "Central American Entomology" |
| " | 467 | " | Physiological Selection and the Origin of Species |
| XXXV. | 366 | 1887 | Mr. Romanes on Physiological Selection |

| | | | |
|---|---|---|---|
| XXXVI. | 530 | " | The British Museum and the American Museums |
| XXXIX. | 611 | 1889 | Which are the Highest Butterflies? (Quotations from Letter of W.H. Edwards) |
| XL. | 619 | " | Lamarck _versus_ Weismann |
| XLI. | 53 | " | Protective Coloration of Eggs |
| XLII. | 289 | 1890 | E.B. Poulton's "Colours of Animals" |
| " | 295 | " | Birds and Flowers |
| XLIII. | 79, 150 | " | Romanes on Physiological Selection |
| " | 337 | 1891 | C. Lloyd Morgan's "Animal Life and Intelligence" |
| " | 396 | " | Remarkable Ancient Sculptures from North-West America |
| XLIV. | 529 | " | David Syme's "Modification of Organisms" |
| XLVI. | 518 | " | Variation and Natural Selection |
| XLV. | 31 | " | Topical Selection and Mimicry |
| " | 553 | 1892 | W.H. Hudson's "The Naturalist in La Plata" |
| XLVI. | 56 | " | Correction in "Island Life" |
| XLVII. | 55 | " | An Ancient Glacial Epoch in Australia |
| " | 175, 227 | " | The Earth's Age |
| " | 437 | 1893 | The Glacial Theory of Alpine Lakes |
| " | 483 | " | W.H. Hudson's "Idle Days in Patagonia |
| XLVIII. | 27 | " | H.O. Forbes's Discoveries in the Chatham Islands |

| "     | 73       | "    | Intelligence of Animals |
|-------|----------|------|--------------------------|
| "     | 198      | "    | The Glacier Theory of Alpine Lakes |
| "     | 267      | "    | The Non-inheritance of Acquired Characters |
| "     | 389      | "    | Pre-natal Influences on Character |
| "     | 390      | "    | Habits of South African Animals |
| "     | 589      | "    | The Supposed Glaciation of Brazil |
| XLIX. | 3        | 1893 | The Recent Glaciation of Tasmania |
| "     | 52, 101  | "    | Sir W. Howorth on "Geology in Nubibus" |
| "     | 53       | "    | Recognition Marks |
| "     | 197, 220 | 1894 | The Origin of Lake Basins |
| "     | 333      | "    | J.H. Stirling's "Darwinianism, Workmen and Work" |
| "     | 549      | "    | B. Kidd's "Social Evolution" |
| "     | 610      | "    | What are Zoological Regions? (Read at Cambridge Natural Science Club) |
| L.    | 196      | "    | Panmixia and Natural Selection |
| "     | 541      | "    | Nature's Method in the Evolution of Life |
| LI.   | 533      | 1895 | Tan Spots over Dogs' Eyes |
| "     | 607      | "    | The Age of the Earth |
| LII.  | 4        | "    | Uniformitarianism in Geology |
| "     | 386      | "    | H. Dyer's "Evolution of Industry" |
| "     | 415      | "    | The Discovery of Natural Selection |
| LIII. | 220      | 1896 | The Cause of an Ice Age |
| "     | 317      | "    | The Astronomical Theory of a Glacial Period |
| "     | 553      | "    | E.D. Cope's "Primary Factors of Organic |

|  |  |  | Evolution" |
|---|---|---|---|
| " | 553 | " | G. Archdall Reid's "Present Evolution of Man" |
| LV. | 289 | 1897 | E.B. Poulton's "Charles Darwin and the Theory |
|  |  |  | of Natural Selection" |
| LIX. | 246 | 1899 | The Utility of Specific Characters |
| LXI. | 273 | 1900 | Is New Zealand a Zoological Region? |
| LXVII. | 296 | 1903 | Genius and the Struggle for Existence |
| LXXV. | 320 | 1907 | Fertilisation of Flowers by Insects |
| LXXVI. | 293 | " | The "Double Drift" Theory of Star Motions |

# Notes

1.

"It is no doubt the chief work of my life."—C. DARWIN.

2.

"My Life," i. 396-7.

3.

"My Life," ii. 94-5.

4.

"My Life," pp. 97-8.

5.

"My Life," pp. 98-9.

6.

Dr. Henry Forbes in a note to the Editor writes: "In his 'Island Life' Wallace extended his philosophical observations to a wider field, and it is in philosophical biology that Wallace's name must stand pre-eminent for all time." "In our own science of biology," say Profs. Geddes and Thomson in a recent work, "we may recall the 'Grand Old Men,' surely second to none in history—Darwin, Wallace, and Hooker."

7.

"My Life," ii. 99-101.

8.

"My Life," ii. 22.

9.

"The Origin of the Races of Man."

10.

"The Malay Archipelago."

11.

"The Malay Archipelago."

12.

Private Secretary to Sir Charles Lyell.

13.

"The Descent of Man."

14.

Probably refers to "The Geographical Distribution of Animals."

15.

The book referred to is Wallace's "Island Life," published in 1880.

16.

For the work on "Darwinism."

17.

Printed in full as a footnote to Weismann's "Essays upon Heredity," etc.

18.

*See* footnote 3, pp. 172-3, of Weismann's "Essays upon Heredity," etc.

19.

"The Origin of Floral Structures through Insect and Other Agencies." Internat. Sci. Series. 1888.

20.

"The Origin of the Fittest." London, 1887.

21.

"Essays upon Heredity and Kindred Biological Problems," Vol. II. 1892.

22.

*Trans. Ent. Soc., London*, 1892, p. 293.

23.

As Hope Professor of Zoology in the University of Oxford.

24.

A member of a family which has produced several eminent medical men.

25.

Vol. I., p. 445, a review of "A Theory of Development and Heredity," by Henry B. Orr. 1893.

26.

"Material for the Study of Variation, treated with especial regard to Discontinuity in the Origin of Species." 1894.

27.

Reprinted in "Essays on Evolution," p. 95. 1908.

28.

"The Present Evolution of Man." 1896.

29.

Presidential Address in Section D of British Association, 1896, reprinted in "Essays on Evolution," p. 1.

30.

To the British Association at Edinburgh, 1892.

31.

Vol. ixx. (1904), p. 313, a review of T.H. Morgan's "Evolution and Adaptation."

32.

"The Bearing of the Study of Insects upon the Question, Are Acquired Characters Hereditary?" The Presidential Address to the Entomological Society of London, 1905, reprinted in "Essays on Evolution," p. 139.

33.

Probably "Root Principles," by Child.

34.

"Essays on Evolution." 1908.

35.

Of the Introduction to "Essays on Evolution."

36.

Vol. lxxvii., p. 54, a note "On the Interpretation of Mendelian Phenomena."

37.

The Oxford Celebration of the Hundredth Anniversary of the Birth of Charles Darwin, February 12, 1809. An account of the celebration is given in "Darwin and 'The Origin,'" by E.B. Poulton, p. 78. 1909.

38.

The Darwin Celebration.

39.

"The World of Life."

40.

*Bedrock*, April, 1912, p. 48.

41.

"Shall we have Common Sense? Some Reeeat Lectures." By George W. Sleeper. Boston, 1849.

42.

*See* footnote to preceding letter. The book formed the subject of Prof. Poulton's Presidential Addresses (May 24, 1913, and May 25, 1914) to the Linnean Society (*Proceedings*, 1912-13, p. 26, and 1913-14, p. 23). The above letter is in part quoted in the former address.

43.

This letter relates to evidences, favourable to Sleeper, which had not at the time been critically examined, but broke down when carefully scrutinised. *See* Prof. Poulton's address to the Linnean Society, May 25, 1914 (*Proc.*, 1913-14, p. 23).

44.

For many years he was Examiner in Physiography at South Kensington.

45.

*See* footnote on p. 109.

46.

For letters from Wallace describing Col. Legge's visit with the Order, *see* pp. 128 and 224.

47.

The present Lord Rothschild.

48.

On his ninetieth birthday.

49.

See his book, "Land Nationalisation, its Necessity and its Aims" (1882).

50.

Although this book was his last published work, it was written before "Social Environment and Moral Progress." He handed me the MS. a few months before his death.—The Editor.

51.

A full account of this scheme is given in his "Studies, Scientific and Social," chap. xxvi.

52.

"My Life," ii. 237-8

53.

Advocating Eugenics and the segregation of the unfit.

54.

Hon. Sec. of the Federated Trades and Labour Council, Bournemouth.

55.

At an Old Age Pension meeting.

56.

*See* Vol. I., p. 20.

57.

"The World of Life," p. 374.

58.

"Life and Letters," i. 58.

59.

Considerable reference is made to Mrs. Hardinge in "Miracles and Modern Spiritualism" pp. 117-21.

60.

The "spirits" are supposed to produce the faces.

61.

This is a strange accompaniment of most advanced spiritual phenomena.

62.

Against vaccination.

63.

Psychical Research Society Report.

64.

"The Wonderful Century."

65.

A medium.

66.

The lecture at the Royal Institution, when he wore the Order.

67.

In *Nature*, Nov. 20, 1913, p. 348.

68.

"The Wonderful Century," p. 437.

69.

"I have been speculating last night," wrote C. Darwin to his son Horace, "what makes a man a discoverer of undiscovered things; and a most perplexing problem it is. Many men who are very clever—much cleverer than the discoverers—never originate anything. As far as I can conjecture, the art consists in habitually searching for the causes and meaning of everything which occurs."—"Emma Darwin," p. 207.

70.

It is interesting to compare this with Darwin's manner of writing. Darwin confessed: "There seems to be a sort of fatality in my mind leading me to put at first my statement or proposition in a wrong or awkward form. Formerly I used to think about my sentences before writing them down; but for several years I have found that it saves time to scribble in a vile hand whole pages as quickly as I possibly can, contracting half the words; and then correct deliberately. Sentences thus scribbled down are often better ones than I could have written deliberately."

71.

See pp. 227, 234.

72.

But see *ante*, p. 153.

73.

Wallace's section of the Darwin-Wallace Essay entitled "On the Tendency of Species to form Varieties; and on the Perpetuation of Varieties and Species by Natural Means of Selection."